Vaccines
Recent Trends and Progress

NATO ASI Series

Advanced Science Institutes Series

A series presenting the results of activities sponsored by the NATO Science Committee, which aims at the dissemination of advanced scientific and technological knowledge, with a view to strengthening links between scientific communities.

The series is published by an international board of publishers in conjunction with the NATO Scientific Affairs Division

A	**Life Sciences**	Plenum Publishing Corporation
B	**Physics**	New York and London
C	**Mathematical and Physical Sciences**	Kluwer Academic Publishers
D	**Behavioral and Social Sciences**	Dordrecht, Boston, and London
E	**Applied Sciences**	
F	**Computer and Systems Sciences**	Springer-Verlag
G	**Ecological Sciences**	Berlin, Heidelberg, New York, London,
H	**Cell Biology**	Paris, Tokyo, Hong Kong, and Barcelona
I	**Global Environmental Change**	

Recent Volumes in this Series

Series A: Life Sciences

Vaccines
Recent Trends and Progress

Edited by

Gregory Gregoriadis

School of Pharmacy
University of London
London, United Kingdom

Anthony C. Allison

Syntex Research
Palo Alto, California

and

George Poste

SmithKline Beecham Pharmaceuticals
King of Prussia, Pennsylvania

Springer Science+Business Media, LLC

Proceedings of a NATO Advanced Study Institute
on Vaccines: Recent Trends and Progress,
held June 24–July 5, 1990,
at Cape Sounion Beach, Greece

Library of Congress Cataloging in Publication Data

NATO Advanced Study Institute on Vaccines: Recent Trends and Progress (1990:
Akra Sounion, Greece)
    Vaccines: recent trends and progress / edited by Gregory Gregoriadis, An-
thony C. Allison, and George Poste.
        p.   cm.—(NATO ASI series. Series A, Life sciences; v. 214)
    "Proceedings of a NATO Advanced Study Institute on Vaccines: Recent Trends
and Progress, held June 24–July 5, 1990, at Cape Sounion Beach, Greece"—T.p.
verso.
    "Published in cooperation with NATO Scientific Affairs Division."
    Includes bibliographical references and index.
    ISBN 978-1-4613-6717-8   ISBN 978-1-4615-3848-6 (eBook)
    DOI 10.1007/978-1-4615-3848-6
    1. Vaccines—Congresses. I. Gregoriadis, Gregory. II. Allison, Anthony C. (An-
thony Clifford), 1925-   . III. Poste, George. IV. North Atlantic Treaty Organization. Scientific Affairs Division. V. Title. VI. Series.
    [DNLM: 1. Vaccines—congresses. QW 805 N279v 1990]
QR189.N37   1990
615'372—dc20
DNLM/DLC   91-29190
for Library of Congress   CIP

ISBN 978-1-4613-6717-8

© 1991 Springer Science+Business Media New York
Originally published by Plenum Press, New York in 1991
Softcover reprint of the hardcover 1st edition 1991

PREFACE

The success of vaccination in controlling infectious diseases is well documented. However, low profitability, expense and liability have hindered research and development of vaccines. Recently, increasing realization (enhanced by the AIDS pandemic) of the need to overcome such difficulties has led to steps being taken by national authorities, non-profit and commercial organizations to resolve them. This has been facilitated by developments in recombinant DNA techniques, the advent of monoclonal anti-bodies and progress in the understanding of the immunological structure of proteins which have laid the foundation of a new generation of vaccines. Such vaccines are defined at the molecular level, can elicit immune responses controlling infectious organisms and are therefore potentially free of the problems encountered in conventional ones. Unfortunately, subunit and synthetic peptide vaccines are often only weakly or non-immunogenic. However, developments in both antigen production and immuno-potentiation of weak antigens have opened new avenues with exciting prospects for vaccine design.

This book contains the proceedings of the 2nd NATO Advanced Studies Institute "Vaccines: Recent Trends and Progress" held in Cape Sounion Beach, Greece, during 24 June-5 July, 1990. It deals with the mechanism of induct-ion of immunity and the role of antigen presenting cells and mediators, a variety of approaches to immunopotentiation including the use of new gener-ation immunological adjuvants, as well as accounts of the experimental and clinical applications of new vaccines. We express our appreciation to Professors K. Dalsgaard, J.H.L. Playfair and H. Snippe for their advice in the planning of the ASI and to Mrs. Janet Abel for her help with the pract-ical aspects of it. We are also grateful to Mrs. Susan Gregoriadis for her invaluable input in the editing of the book. The ASI was held under the sponsorship of NATO Scientific Affairs Division and generously co-sponsored by SmithKline Beecham Pharmaceuticals (Philadelphia). Financial assistance was also provided by Merck and Co. Inc. (West Point), Syntex Research Inc. (Palo Alto), Intervet International B.V. (Boxmeer), Boehringer Mannheim GmbH (Tutzing/Obb), Biogen (Cambridge, MA), SmithKline Biologicals (Rixensart), Connaught Laboratories Ltd. (Willowdale), Superfos Biosector A/S (Vedbaek) and Wyeth Ayerst Research (Philadelphia).

June 1991

Gregory Gregoriadis
Anthony C. Allison
George Poste

CONTENTS

DENDRITIC CELLS IN THE INDUCTION OF IMMUNITY

Stella C. Knight

Division of Immunological Medicine
MRC Clinical Research Centre
Watford Road, Harrow, Middlesex HA1 3UJ, UK

INTRODUCTION

Bone marrow-derived dendritic cells (DC) are potent antigen-presenting cells and are particularly important for their capacity to recruit resting T cells into immune responses (Metlay et al, 1990; Melief 1989). An outline of the life history of DC is shown in Fig. 1. The macrophage and dendritic cell lineages diverge early from a bone-marrow stem cell (Reid et al, 1990) and occasional stem cells (Reid et al, 1990) as well as more mature DC (Steinman, 1989) are also present in peripheral blood. There is evidence that DC may enter the spleen directly and they are also distributed in small numbers to most tissues of the body. The tissue specific stage of their life history is exemplified by the skin Langerhans' cells. In the tissues, the DC appear to act as sentinels of the immune system and, particularly following exposure to antigen, can travel as veiled cells in the afferent lymphatics to the lymph nodes where many may localise in the T dependent, paracortical areas. There is strong circumstantial evidence that they

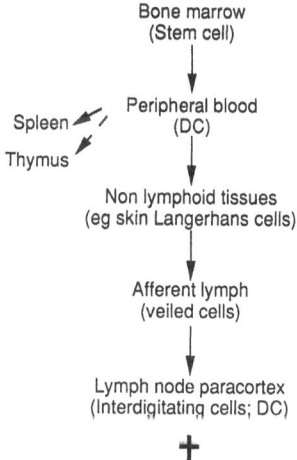

Fig. 1. Dendritic cell life history.

Vaccines, Edited by G. Gregoriadis *et al.*
Plenum Press, New York, 1991

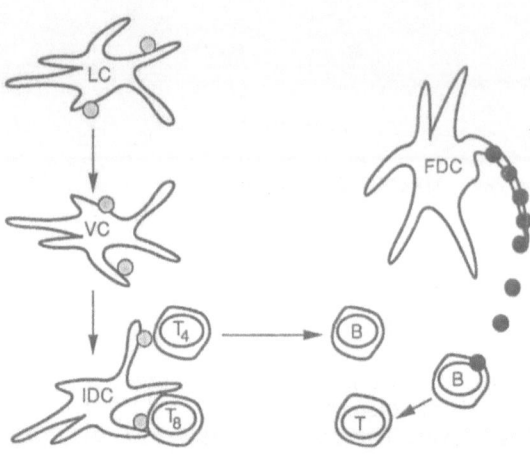

Fig. 2. Bone marrow-derived dendritic cells in peripheral tissues
exemplified by Langerhans' cells (LC) of the skin can acquire
antigen, travel as veiled cells (VC) in the afferent lymphatics
and become interdigitating cells (IDC) of the paracortex of
lymph nodes. These cells may stimulate T4 and T8 cells
directly. Antigen-antibody complexes formed during secondary
exposure can be trapped by follicular dendritic cells (FDC)
and are important in stimulating B cell memory.

become interdigitating cells. It is not clear whether DC from tissues can
re-enter the circulation but there is little evidence of DC leaving the
lymph nodes in afferent lymph. It is believed that they end their life in
the nodes, perhaps through killing by natural killer cells (Shah et al,
1985) or by a specific T cell dependent cytotoxic effect (Macatonia, Taylor,
Askonas and Knight, in preparation).

DC isolated from lymph nodes or spleen are able to cluster T cells non-
specifically. This distinguishes them from antigen-bearing macrophages
which are only able to cluster T cells specifically responding to the antigen. The capacity of DC to cluster T cells non-specifically and then to
retain and activate specific cells is consistent with their peculiar ability
to recruit resting T cells into immune responses. Both T4 and T8 cells can
be activated by DC (Macatonia et al, 1989) (Fig. 2). Once T cells are activated there are many other cells bearing MHC class II molecules that are able
to promote the growth of these pre-activated clones. Within the lymph nodes
the role of follicular dendritic cells (a different cell lineage, probably
derived in situ from fibroblastic elements within the lymph nodes) in memory
responses is established (Tew et al, 1989). These follicular cells are
believed to acquire antigen/antibody complexes and activate B cells which in
turn can promote the T cells responses (Fig. 2). There are suggestions that
activated B cells may be able to initiate resting T cell responses to some
extent (Metlay et al, 1990). However, in our studies of contact sensitizers
and influenza virus in mice we found no evidence that B cells from the lymph
nodes stimulated T cell responses (Tew et al, 1989) and the major route of
stimulation was via the bone-marrow-derived DC.

DC IN INITIATION OF IMMUNE RESPONSES IN VIVO

The capacity of DC to initiate immune responses in vivo is demonstrated from a summary of the events of in vivo sensitization to contact sensitizer shown in Fig. 3 (Macatonia et al, 1989; Knight and Macatonia, 1989). The study used fluorescein isothiocyanate (FITC) which has the dual advantages of being both a potent contact sensitizer and fluorescent, so allowing the identification of cells bearing antigen. Following exposure to FITC there was an increase in numbers of DC in the draining lymph nodes and this preceded any rise in T cell numbers. The DC were the only cells reproducibly carrying high levels of antigen (Fig. 3B). The capacity of these DC carrying antigen to stimulate T cell activity was demonstrated by taking DC from the draining nodes and adding them to normal, naive T cells in vitro. When animals were specific-pathogen-free, normal DC did not cause stimulation of lymphocytes. However, following exposure in vivo to FITC the DC caused syngeneic stimulation (Fig. 3C). This shows the activity of DC in presenting antigen to T cells and also indicates that the basis of the autologous mixed leukocyte reaction may be the presentation of foreign antigens to T cells. The potency of the DC at initiating immunity was also shown from the capacity of small numbers of DC from the draining nodes (e.g. 10^4) to initiate delayed hypersensitivity when injected into naive recipients (Fig. 3D). The capacity of DC to initiate delayed hypersensitivity was seen at the time the DC were increased in number and shown to carry antigen. T cells transferring delayed hypersensitivity developed in the lymph nodes at later times following skin painting (Fig. 3D), presumably in response to the antigen presented on the DC.

PRIMARY PROLIFERATIVE AND CYTOTOXIC T CELL RESPONSES TO VIRAL ANTIGENS IN VITRO

The primary proliferative response of lymphocytes to DC exposed to FITC was demonstrated in Fig. 3C. Such primary proliferative responses can now also be induced in response to more defined antigens (Macatonia et al, 1987). The limitation in producing primary responses to virus previously reported was originally believed to be due to the limitation in the numbers of antigen-specific T cells available from non-primed animals. However, this is not the case because if a more effective antigen-presentation system was combined with an efficient culture technique primary responses were obtained in vitro. Thus, mouse DC exposed in vitro to influenza virus and added to syngeneic T cells in 20 µl hanging drops in vitro initiated primary proliferative and cytotoxic T cell responses (Macatonia et al, 1989). Primary responses to the nucleoprotein peptide known to be the target of in vivo responses to influenza were also obtained (Macatonia et al, 1989).

This primary stimulation system has now been applied to the study of HIV and HIV peptides in vitro. Lymphocytes taken from individuals with no evidence of HIV infection were stimulated in vitro in 20 µl hanging drop cultures by autologous enriched DC populations pulsed with HIV. In all individuals tested, primary proliferative responses and the development of cytotoxic T lymphocytes able to kill virus infected, autologous (but not allogeneic) lymphoblasts were seen (Macatonia, Patterson and Knight, in preparation). This technique could also be used to identify T cell epitopes of the virus. Peptides (15 mers provided via the AIDS Directed Programme in the United Kingdom) were tested as stimulators of primary proliferative and cytotoxic responses (Knight et al, 1990a). One epitope stimulating primary cytotoxic responses that kill virus-infected target cells has been identified (Macatonia, Patterson and Knight, in preparation). This technique for primary stimulation will allow the identification of epitopes stimulatory for human T cells without the necessity for using cells from immunised individuals. Such an approach has applications for identifying potential

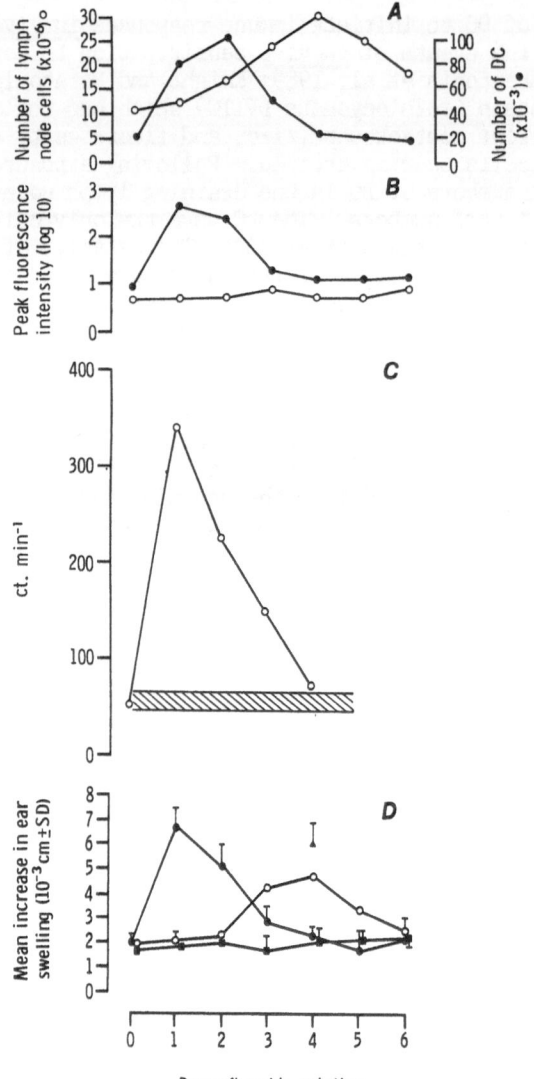

Days after skin painting

Fig. 3. Summary of consequences in lymph nodes of exposure of CBA mice to
a skin contact sensitizer, fluorescein isothiocyanate (for full
experimental details see Knight and Macatonia, 1989 and Macatonia,
Knight et al, 1987). Graphs show: A. Early rise in dendritic cells
in draining lymph nodes followed by a rise in lymphocyte numbers;
B. dendritic cells (●) have high levels of fluorescence indicated
from peak fluorescence intensity measured in isolated cells using
the fluorescence activated cell counter. Other cells show no
expression of antigen (○); C. dendritic cells (500) from mice at
different times after skin painting were added to lymphocytes
(100,000) from naive animals. Cultures were 20 μl hanging drops
and uptake of ^{3}H-thymidine is shown. In specific pathogen free
animals there is no syngeneic stimulation until dendritic cells are
taken from animals exposed to antigen; D. dendritic cells (50,000)
taken from skin painted animals initiated specific delayed hypersensitivity in naive recipient animals (●) and this precedes
capacity of T cells (2 x 10^6 ○ , 10 x 10^6 ▲) to transfer immunity.
Recipient animals were ear-painted 5 days after cell transfer and
ear swelling measured 24 hours later.

4

epitopes for vaccination purposes although appropriate delivery systems for these immunogenic peptides _in vivo_ will be required.

IMMUNOSUPPRESSION BY DENDRITIC CELLS

A further consequence of exposure of DC to HIV can be identified. Normal DC exposed _in vitro_ to HIV became infected with virus (Knight and Macatonia, 1989) and as many as 65% of the DC can contain viral DNA as judged by _in situ_ hybridisation after three days in culture with virus (Knight et al, 1990b). As described in the previous section, such infected cells are able to stimulate primary immune responses to the virus itself. However, their functional capacity in presenting other antigens is blocked (Knight and Macatonia, 1989). High infection of DC and a block in the capacity to present other antigens is also seen using cells from HIV infected patients (Macatonia et al, 1990). The infection and dysfunction in DC is seen in asymptomatic individuals and precedes any changes in T cells. A block in primary recruitment of T cells into immune responses may, therefore, be a fundamental defect in HIV infection. Since stimulation of secondary responses by macrophages is still possible in asymptomatic patients (Macatonia, Pinching and Knight, in preparation) it is likely that there would be no major defect in immunity to common antigens. However, activated T cells may be lost via a variety of mechanisms known to deplete T cells in HIV infection (Macatonia et al, 1990). In the presence of blocked DC recruitment of resting T cells, such losses may become cumulative and lethal.

CONCLUSIONS

Dendritic cells (DC) are potent antigen presenting cells which are vital for the recruitment of resting T cells into immune responses. There is evidence that DC in the periphery (e.g. as Langerhans' cells) can acquire both contact sensitizers or viral antigens, carry them into the lymph nodes and then, as interdigitating cells of the paracortex, cluster T cells and activate cell mediated immune responses. If DC are pulsed with viral antigen _in vitro_ they are also able to initiate specific primary proliferative and cytotoxic T cell responses.

HIV can infect DC _in vitro_ or _in vivo_ and these cells are then blocked in their capacity to present other antigens to T cells. Such a blocked recruitment of resting T cells into immune responses may be a fundamental defect underlying the gradual and cumulative depletion of T cells seen in AIDS.

However, DC from normal, non HIV-infected individuals can be exposed to HIV or peptides of HIV _in vitro_. These will stimulate primary proliferative and cytotoxic responses in autologous T cells _in vitro_. This _in vitro_ primary stimulation system thus allows the direct identification of _human_ T cell epitopes of HIV by short term lymphocyte stimulation studies using non-immune donors and will, therefore, be useful for identifying material with potential as vaccines. The success in raising cytotoxic T cells that kill virus-infected target cells in primary responses _in vitro_ using small peptides, provides strong motivation for identifying an appropriate route of delivery for such small peptides to produce immunity _in vivo._

REFERENCES

Knight, S.C. and Macatonia, S.E., 1989, Dendritic cells and T cells transfer sensitization for delayed-type hypersensitivity after skin painting with contact sensitizer, _Immunology_, 66:99.

Knight, S.C., Macatonia, S.E., Bedford, P.A. and Patterson, S., 1990a, Dendritic cells and HIV infection, in: "Morphological and Functional Aspects of Accessory Cells in Retroviral Infection", P. Racz, ed., Karger, Basel, in press.

Knight, S.C., Macatonia, S.E. and Patterson, S., 1990b, HIV-I infection of dendritic cells, in: "International Reviews of Immunology", R.M. Steinman, ed., Harwood Academic Publishers, Switzerland, in press.

Macatonia, S.E., Knight, S.C., Edwards, A.J., Griffiths, S. and Fryer, P., 1987, Localization of antigen on lymph node dendritic cells after exposure to the contact sensitizer fluorescein isothiocyanate. Functional and morphological studies, J.exp.Med., 166:1654.

Macatonia, S.E., Taylor, P., Knight, S.C. and Askonas, B., 1989, Primary stimulation by dendritic cells induces antiviral proliferative and cytotoxic cell responses in vitro, J.exp.Med., 269:1255.

Macatonia, S.E., Lau, R., Patterson, S., Pinching, A.J. and Knight, S.C., 1990, Dendritic cell depletion and dysfunction in HIV infected patients, Immunology, in press.

Melief, C.J.M., 1989, Dendritic cells as specialized antigen-presenting cells, in: "Lymphoid dendritic cells: Their life history and roles in immune responses", Res.Immunol., 140:902.

Metlay, J.P., Pure, E. and Steinman, R.M., 1990, Control of immune response at the level of antigen-presenting cells: A comparison of the function of dendritic cells and B lymphocytes, Adv.Immunol. 47:45.

Reid, C.D.L., Fryer, P.R., Clifford, C., Kirk, A. and Knight, S.C., 1990, Haematopoietic progenitors common to dendritic/Langerhans' cells and macrophages circulate in high numbers in the blood of leukaemia and lymphoma patients recovering from treatment induced aplasia, Blood.

Shah, P.D., Gilbertson, S.M. and Rowley, D.A., 1985, Dendritic cells that have interacted with antigen are targets for natural killer cells, J.exp.Med., 162:625.

Steinman, R.M., 1989, Dendritic cells: Clinical aspects, Res.Immunol., 140:911.

Tew, J.G., Kosco, M.H. and Szakal, A.K., 1989, The alternative antigen pathway, Immunol.Today, 10:229.

6

ESCAPE MECHANISMS OF VIRUSES FROM IMMUNE RESPONSES AND THEIR RELEVANCE TO

VACCINE DESIGN

Jean-Louis Virelizier

Unite d'Immunologie Virale
Institut Pasteur 75724
Paris, Cedex 15, France

INTRODUCTION

To survive as a species, each virus family has had to use a particular
strategy of replication and dissemination adapted to both the intrinsic
properties of the viral genome and the abilities of the host's immune system
to mount efficient immune responses aimed at eradicating the virus from the
body. It is likely that during evolution the virus genomes which have
survived have done so because of either versatility permitting antigenic
variation, or an ability to integrate into the host genome without trans-
cribing viral genes or else a capacity to take advantage for their replication of transcriptional mechanisms which immunocompetent cells use for
their own activation. The present communication is not intended to exhaust-
ively review the varied situations where viruses survive immune attack in
single hosts or in whole populations, but rather to discuss a few examples
of host-virus relationships suggestive of general strategies of viruses to
cope with their host's environment, to remain infective and to ensure their
transmission from one individual to another. Instead of the usual description of the arsenal of host defense mechanisms, we propose a reflection
about failures of the immune system to eradicate virus diseases, with the
hope that such a salubrious exercise will help in designing better, future
vaccines.

ANTIGENIC VARIATION AS AN ESCAPE STRATEGY

The more efficient an immune response mechanism, the more appropriate
and adapted must be the viral escape strategy involved. The well-documented
antigenic drift of influenza virus provides an example of how a virus genome
can escape specific humoral, and possibly cell-mediated, immune mechanisms.
The main mechanism of prevention and recovery from influenza infection
appears to be the production of antibody to the hemagglutinin (HA). In the
experimental mouse model, it has been shown that passive transfer of antibody to HA prevents infection and permits recovery, even when administration
of antibody is performed after infection. The specificity of transferred
antibodies is critical. Transfer of antibodies to antigenic determinants
common to all influenza strains along a given influenza subtype is not protective. In contrast, transfer of antibody to the strain-specific determinants of the influenza HA is highly protective, even in immunosuppressed
mice unable to mount specific immune responses, or when used in such low

amounts as to be undetectable after dilution in recipient animal sera by sensitive antibody detection techniques in the serum. This indicates that the specific antibody response to strain-specific determinants of HA is sufficient to prevent infection and eradicate influenza virus from the experimentally infected host (Virelizier, 1975). The situation in humans is likely to be more complex than the simple experimental model described above because antibody to neuraminidase is likely to participate in protection, and influenza strains of more than a given serotype may coexist in human populations. It is quite likely, however, that the cross-reactive antibody to HA present in a human population immunized by a previous epidemic strain is not efficient at preventing infection with a new strain of influenza virus. The situation is such that influenza virus, by modifying critical neutralization epitopes, may initiate a new epidemic. In each infected individual, the virus is eradicated within a few days by the appearance of strain-specific, protective antibody. The necessary delay in mounting such response in the infected host makes it possible for the virus to be trans-mitted, essentially by sneezing, to other susceptible individuals. A new epidemic can thus spread in the population. Such an escape strategy implies that influenza virus replicates rapidly in primary target cells of the res-piratory tract, which is indeed the case. A slowly replicating viral genome could not use this strategy.

T lymphocytes also participate in host recovery from influenza virus infection. They do so through their helper effect, by permitting the generation of an optimal antibody response by B lymphocytes, and possibly also through their cytotoxic potential. Cytotoxic T cells (CTL) have been shown to specifically recognize influenza viral antigens. It has been shown that "internal" influenza antigens, especially the nucleoprotein (NP), are expressed on the membrane of influenza-infected cells (Virelizier et al, 1977), and are recognized in an "immunodominant" way (i.e. preferentially) by CTL (Townsend and Skehel, 1984). This provides a potential mechanism of host defense specific for antigens which are relatively conserved among many influenza subtypes during antigenic drift. However, NP also varies anti-genically, although less frequently than HA. NP variation during antigenic drift is another potential escape mechanism of influenza virus. Indeed, target cells infected with influenza virus genomes coding for NP protein with significant antigenic variation within a given influenza sybtype are not any longer recognized and lysed by CTL (Townsend and Skehel, 1984). If indeed CTL have a role in recovery from influenza infection, such antigenic variation of NP antigen could be yet another way for the virus to escape immunity. This reasoning is in keeping with recent observations made in the experimental lymphocytic choriomeningitis virus (LCMV) model (Pircher et al, 1990), showing that pre-establishment in transgenic mice of specific T cell recognition favors the emergence of viral variants no longer recognized by CTL because of mutation in the relevant T cell epitope.

It should be stressed, however, that helper T lymphocytes also recog-nize internal, relatively conserved antigens, such as nucleoprotein (NP) or matrix (M) protein. Helper T cells specific for NP or M have been shown to help in vivo anti-HA antibody responses (Scherle and Gerhard, 1986). It may thus be of interest to use NP or M protein, together with HA in future influenza vaccines, in the hope of getting earlier and more efficient anti-body responses to HA through the "cross-help" phenomenon provided by helper T cells recognizing conserved influenza epitopes in newly occurring viral strains. It could be argued that helper T cell epitopes, demonstrably present within the HA molecule, suffice to provide cross-help to B cells producing new strain-specific, protective antibody. This is likely to occur during antigenic drift, but not when a new recombinant virus with a totally different HA (antigenic shift) appears, as happened for example in 1968 (H3 N2 strains).

We have discussed above that the production of non-neutralizing antibody may be inefficient. In the case of arboviruses, especially flaviviruses, inappropriate antibody responses may in fact be deleterious for the host. This is clearly the case in vitro with dengue (Halstead et al, 1984) and yellow fever (Schlesinger and Brandriss, 1983) infection. Through the phenomenon of "immune enhancement", flaviviruses subvert the opsonization process and use non-neutralizing antibody to enter permissive Fc-γ receptor-positive cells such as macrophages more easily. The virus-antibody immune complexes, far from neutralizing infection, remain infective. A "rolling snow ball" phenomenon is thus initiated. The more antibody present before infection (cross-reactive antibodies induced by previous flavivirus infection with a different serotype), the more infectious immune complexes are made, the more virus replication will be observed in macrophages, and the more immune complexes will be generated. This is likely to lead to increased virus load and release of toxic monokines by activated macrophages, resulting in enhanced shock and hemorrhagic syndrome, as observed during flavivirus infections of individuals previously immunized with a different serotype of the same virus.

Anti-flavivirus vaccines may have to take into account this phenomenon. The answer to this problem may be either to vaccinate with "neutralizing epitopes" from many serotype-specific strains of a given flavivirus group, or to include in future vaccine preparations flavivirus epitopes conserved among most serotypes. Interestingly, it has been shown that transfer of antibody against non-structural proteins, or active immunization with such antigens conserved among serotypes, protects mice against challenge (Schlesinger et al, 1985). Although such protection is at best partial, this argues in favor of including in anti-flavivirus vaccines conserved, non-structural antigens in addition to serotype-specific antigenic domains. Much remains to be learned about the specificities, classes and sub-classes of antibodies that either protect the host or enhance virus replication in vivo. It would also be of great interest to know whether T cells capable of providing "cross-help" are induced during flavivirus infections. If so, future anti-flavivirus vaccine preparations including non-structural epitopes would have the advantage of inducing both antibody to conserved antigens unable to participate in immune enhancement while providing some protection, and helper T cell memory which might permit an earlier neutralizing antibody response to newly encountered flavivirus strains.

PERPETUATION OF LENTIVIRUS REPLICATION IN MONOCYTE-MACROPHAGES

Lentiviruses are non-oncogenic retroviruses that cause persistent infections with slowly progressive diseases in sheep (visna), goats (CAEV) and horses (EAIV), and acquired immune deficiency in apes (SIV), cats (FIV), cattle (BIV) and man (HIV). Visna virus replication in macrophages has been specially well documented. During the course of visna infection, antigenic variants arise in vivo. These variants react very poorly with antibodies generated in response to the original viral inoculum. This phenomenon was originally interpreted as yet another example of virus escape through antigenic variation induced by the selection pressure exerted by antibody. More recent findings, however, impose reconsideration of such an interpretation. Much depends on the cell type used for the antibody neutralization assay (Kennedy-Stoskopf and Narayan, 1986). In fibroblasts, the site of virus neutralization is at the cell membrane, so that antibody prevents virus attachment. In contrast, the kinetics of visna virus "neutralization" in macrophage cultures is much slower. The requirement for prolonged incubation (15 minutes at 37°C) of virus and antibody before neutralization occurs indicates that the interaction with antibody is initially reversible due to the low affinity of antibodies. Visna virus shows a greater affinity for macrophages than does the antibody towards the virus. Furthermore non-neutralizable mutants of visna virus fail to replace the neutralizable parental virus in immune sheep in which antigenic drift of the virus has occurred (Lutley et al, 1983). Globally, it is now thought that lenti-viruses persist irrespective of whether the animal fails to develop neutra-lizing antibody, as observed in goats infected with CAEV, or do develop such antibodies, as in visna virus infection of sheep. It appears that lenti-viruses have developed such a strong tropism for monocytic target cells that they can overcome the neutralizing antibody response of the host. The host-lentivirus relationship seems to be regulated by the ability of individual cells of the monocyte-macrophage lineage to support minimal, but persistent virus replication, itself modulated by the maturation stage of the infected cells. In this respect, the bone marrow may be a major reservoir of infect-ed cells that do not become involved in pathologic processes until they leave the bone marrow and differentiate into tissue macrophages (Gendelman et al, 1985).

HIV COMBINES A STRATEGY OF LATENCY/REACTIVATION IN T CELLS AND OF PERSISTENT REPLICATION IN MONOCYTES

Among lentiviruses, the agents of AIDS in man and other species are peculiar in that they are able to infect, in addition to monocyte-macro-phages, cells of the T cell lineage. Both cell types bear the membrane CD4 molecule, which can bind avidly to the envelope protein of HIV and SIV. The replication strategy of these viruses, however, appears to be quite distinct in these two cell types.

In T lymphocytes, few HIV proviruses are integrated, and they remain essentially in a quiescent state as long as the cell stays in its normal resting state. Intense HIV replication is observed in T cells activated in vitro. This suggests that transcriptional mechanisms involved in gene ex-pression during T cell activation are used by the virus for its own replic-ation, as discussed previously (Virelizier, 1990). The regulatory region (LTR) of HIV contains an enhancer sequence (Nabel and Baltimore, 1987) that initiates HIV genome transcription after specific binding of the transcrip-tion factor NF-kB induced in lymphoblastoid cell lines by phorbol esters (PMA). We and others have found that tumor necrosis factor (TNF), a natural cytokine induced by specific T cell responses and non-specific inflammatory processes, can induce NF-kB translocation from the cytoplasm to the nucleus, thus initiating HIV transcription. However this phenomenon, observed in lymphoblastoid T cell lines, is unlikely to be relevant to HIV reactivation in resting T lymphocytes, since such cells do not bear detectable TNF recep-tors. We have thus recently embarked on a systematic analysis of the immun-ological signals which induce HIV LTR transactivation in normal, IL2-depend-ent T cell clones. These cells do express TNF receptors, and recombinant TNF indeed induces NF-kB translocation in such cultures. We found, however, that NF-kB translocation is not sufficient to induce HIV LTR translocation in T cell clones. Only multiple transmembrane signals provided by antigen presentation on the membrane of accessory cells (also mimicked by PMA) were able to induce HIV LTR transactivation in normal T cells (Hazan et al, 1990). This indicates that specific antigen recognition by the T cell receptor-CD3 complex, in association with accessory transmembrane molecules, is likely to be the main trigger of HIV reactivation in resting T lympho-cytes. The paradox is that T lymphocytes, of which activation is critically needed for host defense against infectious agent, may be the ideal environ-ment for HIV persistence and replication. In resting T cells, HIV finds a suitable site for latency. In infected T cells activated by antigens, HIV can replicate intensely. This is compatible with the concept that HIV infection not only escapes immunological effector mechanisms, as do other lentiviruses, but also uses immune responses to its own advantage. As far as T lymphocytes are concerned, the infected host is offered the alternative of either not mounting immune responses, with consequential opportunistic infections, or activating infected, immunocompetent cells, thus triggering HIV replication. It is likely that future vaccines against HIV will have to act before such a vicious circle is initiated, and thus before virus dissemination has occurred, ideally preventing virus entry into the body.

Unfortunately, HIV uses a parallel and apparently different strategy of replication in tissue macrophages (Meltzer et al, 1990). Cells of the myelo-monocytic lineage (microglial cells) of the brain show persistent transcription of the HIV genome. Such cells, scattered in various parts of the brain, are not associated with granuloma formation, so that permanent activation of these cells need not be postulated. PCR examination has recently shown that most HIV DNA in brain tissue is unintegrated (Pang et al, 1990). Present results in our laboratory (Bachelerie et al, in preparation) indicate that activity of the HIV enhancer is increased in monocytes during chronic HIV infection. This phenomenon, in addition to the amplifying effect of HIV tat production, is likely to permit perpetuation of HIV transcription in infected monocytes. Thus HIV appears to be able to induce permissiveness of monocytes to its own replication, a phenomenon that may ensure chronic HIV infection in the absence of exogenous activation signals. Once established, HIV infection in monocytes may be self-perpetuating. This is an additional incentive to design anti-HIV vaccines which can prevent early dissemination of the virus to the myelo-monocytic cell lineage.

REFERENCES

Gendelman, H.E., Narayan, O., Molineaux, S., Clements, J.E. and Ghotbi, Z., 1985, Slow, persistent replication of lentiviruses: role of tissue macrophages and macrophage precursors in bone marrow, Proc.Natl.Acad.Sci.USA, 82:7086.

Halstead, S.B., Venkrateshan, C.N., Gentry, M.K. and Larsen, L.K., 1984, Heterogeneity of infection enhancement of dengue 2 strains by monoclonal antibodies, J.Immunol., 132:1529.

Hazan, U., Thomas, D., Alcami, J., Bachelerie, F., Israel, N., Yssel, H., Virelizier, J-L. and Arenzana-Seisdedos, F., 1990, Human T cell clone stimulation with anti -CD3 or tumor necrosis factor induces NF-kB translocation but not human immunodeficiency enhancer-dependent transcription, Proc.Natl.Acad.Sci.USA, in press.

Kennedy-Stoskopf, S. and Narayan, O., 1986, Neutralizing antibodies to visna lentivirus: mechanism of action and possible role in virus persistence, J.Virol., 59:37.

Lutley, R., Petursson, G., Palsson, P.A., Georgsson, G., Klein, J. and Nathanson, N., 1983, Antigenic drift in visna: virus variation during long-term infection of Icelandic sheep, J.Gen.Virol., 64:1433.

Meltzer, M.S., Skillman, D.R., Hoover, D.L., Hanson, B.D., Turpin, J.A., Kalter, D.C. and Gendelman, H.E., 1990, Macrophages and the immunodeficiency virus, Immunol.Today, 11:217.

Nabel, G. and Baltimore, D., 1987, An inducible transcription factor activates expressions of human immunodeficiency virus in T cells, Nature, 326:711.

Pang, S., Koyanagi, Y., Miles, S., Niley, C., Vinters, H.V. and Chen, I.S.Y., 1990, High levels of unintegrated HIV-1 DNA in brain tissue of AIDS dementia patients, Nature, 343:85.

Pircher, H., Moskophidis, D., Rohrer, V., Burki, K., Hengartner, H. and Zinkernagel, R., 1990, Viral escape by selection of cytotoxic T cell-resistant virus variants in vivo, Nature, 346:629.

Scherle, P.A. and Gerhard, W., 1986, Functional analysis of influenza-specific helper T cell clones in vivo. T cells specific for internal viral proteins provide cognate help for B cell responses to hemagglutinin, J.Exp.Med., 164:1114.

Schlesinger, J.J. and Brandriss, M.W., 1983, 17 D yellow fever virus infection of P388D1 cells mediated by monoclonal antibodies: properties of the macrophage Fc receptor, J.Gen.Virol., 64:255.

Schlesinger, J.J., Brandriss, M.W. and Walsh, E.E., 1985, Protection against 17 D yellow fever encephalitis in mice by passive transfer of monoclonal antibodies to the non-structural glycoprotein gp 48 and by active immunization with gp 48, J.Immunol., 135:2805.

Townsend, A.R.M. and Skehel, J.J., 1984, The influenza A virus nucleoprotein gene controls the induction of both subtype specific and cross-reactive cytotoxic T cells, J.Exp.Med., 160:532.

Virelizier, J-L., 1975, Host defenses against influenza virus: the role of anti hemagglutinin antibody, J.Immunol., 115:434.

Virelizier, J-L., Allison, A.C., Oxford, J.S. and Schild, G.C., 1977, Early presence of ribonucleoprotein antigen on surface of influenza virus-infected cells, Nature, 266:52.

Virelizier, J-L., 1990, Cellular activation and human immunodeficiency virus infection, Curr.Opin.in Immunol., 2:409.

ENHANCED IMMUNOGENICITY OF RECOMBINANT AND SYNTHETIC PEPTIDE VACCINES

M.J. Francis

Wellcome Research Laboratories
Langley Court
Beckenham, Kent BR3 3BS, UK

INTRODUCTION

In an attempt to produce more stable and defined vaccines, scientists have been studying the immune responses to many infectious agents in detail in order to identify the critical epitopes involved in providing protective immunity. Armed with this knowledge, it is now possible to mimic such epitopes by producing short peptides and to use these as the basis of a vaccine (Francis, 1990). However, once a vaccine candidate peptide has been identified or predicted then it must be delivered to the immune system in a suitable manner in order to elicit not just a high titre anti-peptide response but anti-peptide antibodies that will recognise and neutralize the infectious agent. Indeed, there has been a widely held view that, due to their relatively small molecular size, peptides are necessarily poor immunogens and thus require carrier-coupling to enhance their immunogenicity. As a result of this there are many examples of elegantly defined peptides which, having been coupled in an uncontrolled manner to large undefined carrier proteins, produced anti-peptide antibodies that totally failed to recognise the native protein.

The presently held concept that peptides behave like haptens is in many cases misguided. Experiments using a 20 amino acid peptide from foot-and-mouth disease virus (FMDV) (Bittle et al, 1982; Pfaff et al, 1982) have demonstrated that the role of keyhole limpet haemocyanin (KLH) as a carrier in priming for a peptide response is fundamentally different from its role in carrier priming since an uncoupled peptide or peptide coupled to a different carrier (tetanus toxoid) could boost a response in peptide-KLH primed animals (Francis et al, 1985). This observation has led to the demonstration of helper T-cell and B-cell determinants on this relatively small peptide (Francis et al, 1987a). Indeed, it is now clear that uncoupled peptides can be immunogenic provided they contain appropriate antibody recognition sites (B-cell epitopes) as well as sites capable of eliciting T-cell help for antibody production (Th-cell epitopes). These Th-cell epitopes must interact with class II major histocompatibility complex (MHC) molecules on the surface of antigen presenting cells (APC) and B-cells and subsequently bind to a T-cell receptor in the form of a trimolecular complex. The Th-cells will provide signals in the form of chemical messengers (lymphokines) to specific B-cells which result in differentiation, proliferation and antibody production. With this knowledge, synthetic peptides can be constructed with appropriate sites for antibody production plus

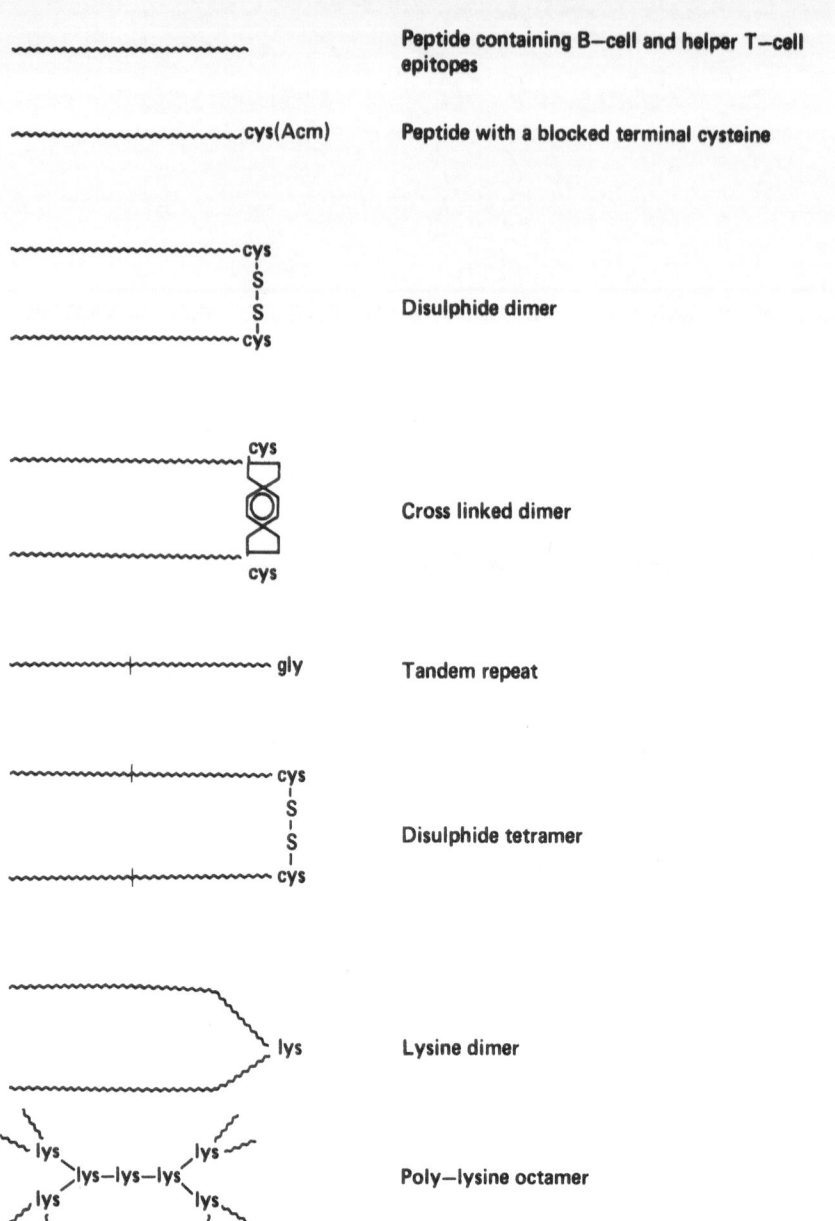

Fig. 1.   Various forms of uncoupled synthetic peptides used for
comparative immunogenicity studies.

additional T-cell epitopes (Francis and Clarke, 1989). This report
describes the polymeric presentation of such peptides as fully synthetic
products and compares their immunogenicity to that of defined peptide fusion
proteins produced by recombinant DNA technology.

SYNTHETIC PEPTIDE PRESENTATION

For this study the 141-160 peptide from VP1 of FMDV was chosen since it
was known to be immunogenic in the absence of a carrier protein if it was
polymerised either with glutaraldehyde or air-oxidised after adding a cysteine residue on each terminus (Bittle et al, 1984). The free peptide was

also immunogenic when delivered in small unilamellar liposomes (Francis et al, 1985). Therefore a number of synthetic forms were prepared (Fig. 1) and subsequently tested for their immunogenicity.

Disulphide Dimers

In the first of these studies the role of a non-natural terminal cysteine residue was examined by comparing the immunogenicity of the 141-160 peptide, which contains B- and Th-cell epitopes; with 141-160 peptide plus an additional C-terminal cysteine residue (141-160cys) and 141-160cys with an S-acetamidomethyl thiol blocking group on the cysteine (141-160cys[Acm]) using incomplete Freund's adjuvant (IFA) (Francis et al, 1987a). No primary neutralizing antibody response was seen following a 100 ug dose of the 141-160 peptide. However, the addition of a non-natural unblocked cysteine residue at the C-terminus resulted in a primary neutralizing antibody response of >2.0 $\log_{10}$ SN_{50}. If the thiol group on the cysteine was blocked as in the 141-160cys (Acm) peptide no such response was seen. Thus the presence of a C-terminal cysteine with a free thiol group greatly enhances the immunogenicity of free 141-160 peptide in IFA. A similar improvement in immunogenicity of free FMDV peptide with added single or multiple cysteine residues has been reported previously (Bittle et al, 1984; Di Marchi et al, 1986).

The presence of a free thiol cysteine residue is likely to result in the formation of peptide dimers that may have a more ordered secondary structure, cause cross-linking of B-cell receptors and/or lead to immune complex formation in vivo. To test whether a fixed dimer form of the peptide was immunogenic, we cross linked the monomer 141-160cys peptides with N,N'-1,4-phenylenediamaleimide via their free sulphydryls and HPLC purified the resultant dimer peptides (McGinn and Francis, 1986, unpublished data). The immune response of guinea pigs to these cross-linked dimers was indistinguishable from that obtained with air-oxidised disulphide dimers using a 100 ug dose of each peptide in IFA (Fig. 2). Therefore, we concluded that the dimer form of the peptide was responsible for its increased immunogenicity and not simply the presence of a free sulphydryl group.

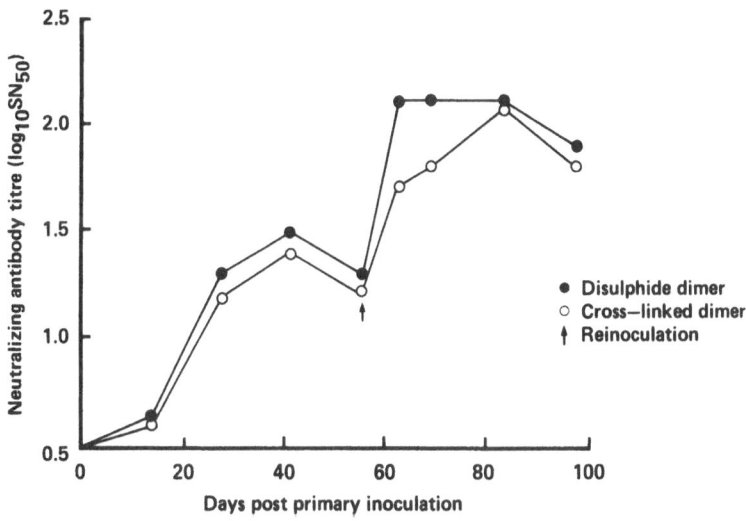

Fig. 2. Neutralizing antibody response of guinea pigs inoculated with disulphide dimers and cross-linked dimers of FMDV VP1 141-160 peptide.

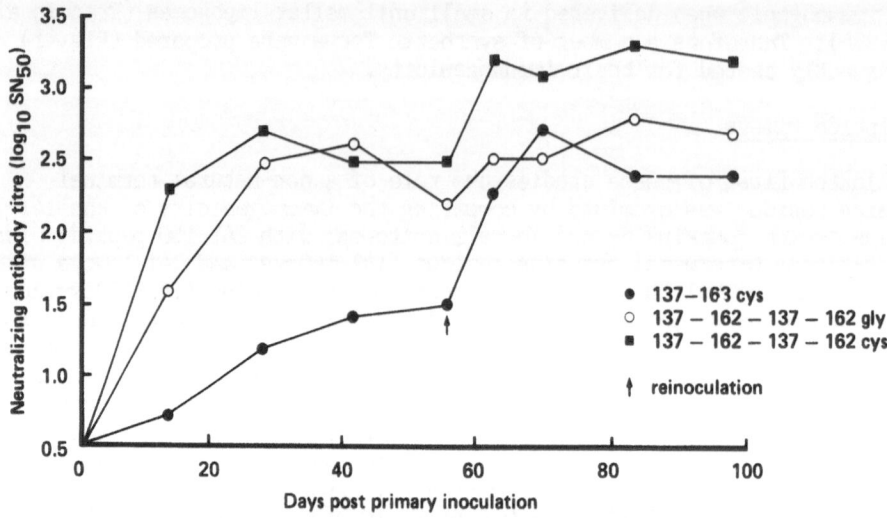

Fig. 3. Neutralizing antibody response of guinea pigs inoculated
with tandem repeats of FMDV VP1 137-162 peptide with or
without a C-terminal cysteine residue.

Tandem Repeats

Following the observation that FMDV peptide disulphide dimers were more
immunogenic than monomers, dimer peptides in the form of tandem repeats were
synthesised with or without a C-terminal cysteine residue. The immunogenicity of these tandem repeats (137-162(x2)cys and 137-162(x2)gly) was compared to that of a single copy 137-163cys peptide using a 150 μg dose in
IFA. The results of this experiment (Fig. 3) demonstrated that tandem
repeats of the FMDV peptide were generally more immunogenic than single copy
disulphide dimers and that the addition of a cysteine residue which could
result in the formation of disulphide tetramer structures may also improve
the response further. The enhanced activity of similar tandem repeats peptides from single serotypes of FMDV as well as double serotypes (O and A)
has been reported previously (Francis et al, 1990a).

Multiple Antigenic Peptide (MAP) System

In order to fully exploit the concept of multiple copy synethetic peptides we have studied the MAP system (Tam, 1988; Posnett et al, 1988) in
collaboration with Dr. J.P. Tam of the Rockefeller University, New York.
This system provides a method for direct solid phase synthesis of a peptide
antigen on to a branching lysine backbone and has been used to produce
several poly-lysine octamer constructs (Tam, 1989). Neutralizing antibody
responses known to protect guinea pigs against challenge infection (Francis
et al, 1988) were obtained following a single inoculation of 0.8 to 4 μg of
peptide in IFA, presented as an octamer (Fig. 4) or a tetramer, whereas 20
μg of the lysine dimer were required to produce a similar level of antibody
(Francis et al, 1991). A monomeric preparation did not evoke measurable
levels of neutralizing antibody at doses up to 20 μg. On a weight for
weight basis octomer peptides appear to be 25-50 fold more immunogenic than
disulphide dimers. Another interesting feature of the MAPs is their ability
to elicit significant levels of neutralizing antibodies following two inoculations with aluminium hydroxide (Francis et al, 1991), the only adjuvant
currently licenced for use in man. FMDV peptides have previously been shown
to be poorly immunogenic when inoculated with this adjuvant (Francis et al,
1987a) unless they were coupled to a protein carrier (Bittle et al, 1982;

16

Francis et al, 1985). Therefore, in this respect the octameric MAP is behaving like a carrier linked peptide, possibly due to its improved adsorption to Al(OH)$_3$ when compared to that of the 141–160cys disulphide dimer (M.J. Francis, unpublished observation).

RECOMBINANT PEPTIDE PRESENTATION

The requirement for multiple copy presentation has also been investigated using recombinant DNA technology by fusing small peptide sequences to the genes coding for larger proteins in order to produce a number of novel constructs (Fig. 5).

β–Galactosidase Fusion Proteins

The use of peptide sequences fused to bacterial proteins as immunogens has the potential advantage of a completely uniform and defined structure compared with the uncharacterised and variable nature of peptide/carrier conjugates prepared by chemical cross-linking. This approach has been used independently by two groups (Broekhuijsen et al, 1986a; Winther et al, 1986) to express FMDV peptides fused to the N-terminus of β–galactosidase in E. coli cells. β–Galactosidase was chosen because it had been shown that antibodies can be produced to foreign proteins located at the N-terminus and it has been shown to contain a number of helper T-cell sites (Krzych et al, 1982; Manca et al, 1985). Preliminary experiments with β–galactosidase (Broekhuijsen et al, 1986b) and TrpLE (Kleid et al, 1985) fusion proteins had indicated that multiple copies of the inserted peptide sequence may be beneficial. Therefore, in collaboration with Dr. B.E. Enger-Valk's group from Medical Biological Laboratory TNO, The Netherlands, we have examined the immunogenicity of one, two or four copies of FMDV VP1 peptide 137 to 162 fused to the N-terminus of β–galactosidase both in laboratory animals and

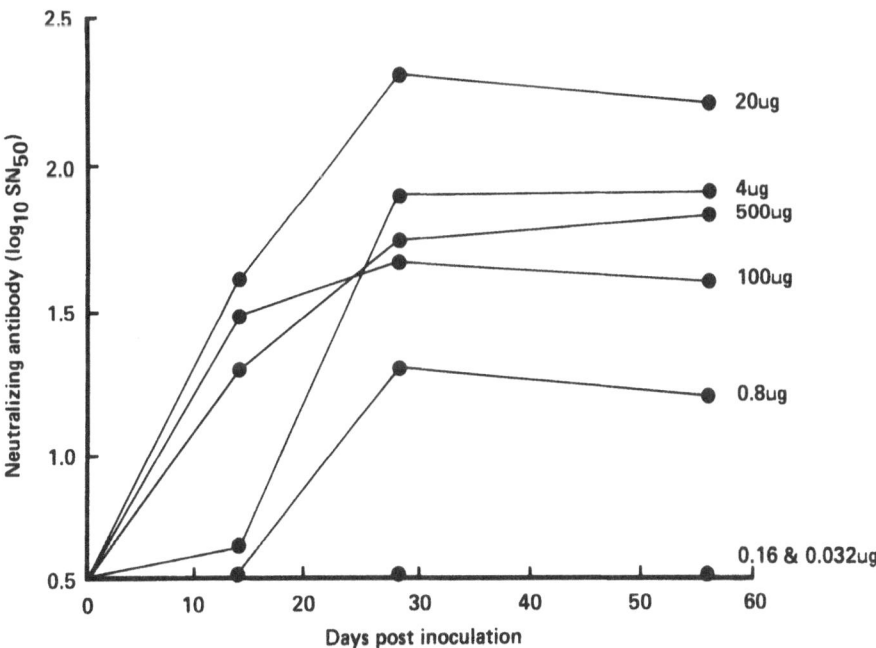

Fig. 4. Neutralizing antibody response of guinea pigs inoculated with octameric FMDV VP1 141–160 MAP at various doses ranging from 500 μg to 0.032 μg.

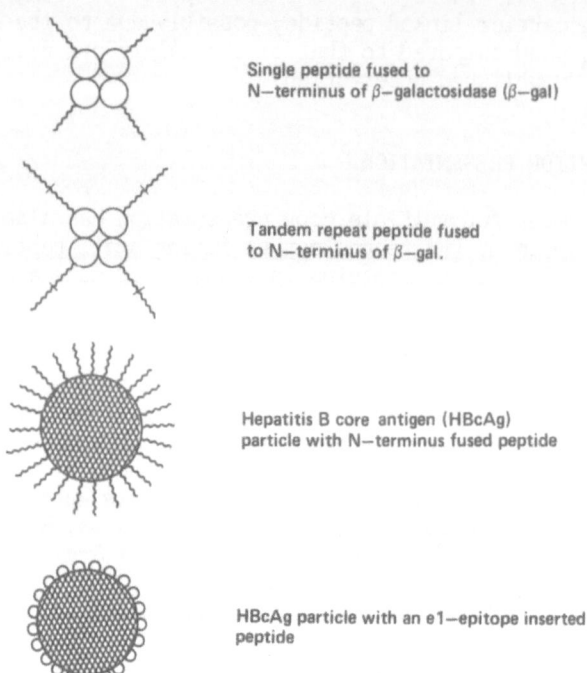

Single peptide fused to
N—terminus of β—galactosidase (β—gal)

Tandem repeat peptide fused
to N—terminus of β—gal.

Hepatitis B core antigen (HBcAg)
particle with N—terminus fused peptide

HBcAg particle with an e1—epitope inserted
peptide

Fig. 5. Various forms of recombinant peptide/protein fusion molecules used for comparative immunogenicity studies.

target species. The protein containing one copy of the viral determinant elicited only low levels of neutralizing antibody whereas protective levels were elicited by proteins containing two or four copies of the determinant (Broekhuijsen et al, 1987). Furthermore, single inoculations of the two copy and four copy proteins, containing as little as 2 µg or 0.8 µg of peptide respectively, were sufficient to protect all the animals against challenge infection. The equivalent of 40 µg of peptide in the four copy protein also protected pigs against challenge infection after one inoculation. Thus the immunogenicity of the multiple copy peptide/β-galactosidase fusion proteins is similar to that obtained using the synthetic MAP system.

Hepatitis B Core Antigen Particles

More recently a further development of the fusion protein concept for multiple peptide presentation has led to the production of particulate structures with epitopes repeated over their entire surface. To date work has principally concentrated on three proteins, hepatitis B surface antigen (Delpeyroux et al, 1986), hepatitis B core antigen (HBcAg) (Newton et al, 1987; Clarke et al, 1987) and yeast Ty protein (Adams et al, 1987), which spontaneously self assemble into 22, 27 and 60 nm particles respectively.

We have shown using HBcAg fusion particles (CFPs) that the immunogenicity of FMDV peptide can approach that of the inactivated virus. Indeed, as little as 0.2 µg of FMDV VP1 142-160 peptide corresponding to 10% of the fusion protein, presented on the surface of CFPs gave full protection to guinea pigs (Clarke et al, 1987). In subsequent experiments N-terminal CFPs were shown to be one hundred-fold more immunogenic than free disulphide dimer synthetic peptides containing B- and T-cell determinants and ten-fold more immunogenic than carrier linked peptide (Francis et al, 1990b). This activity appears to be dependent both on the provision of T-cell help from the HBcAg (Millich et al, 1987), and on particle formation (Clarke et al, 1990). CFPs are also active with or without medically acceptable adjuvants in a wide range of species. Furthermore, systemic responses can be elicited by oral or nasal administration and in a T-cell independent manner (Francis et al, 1990b). This last property of the CFPs offers the possibility of developing vaccine-based therapies for immunocompromised individuals infected with HIV.

In addition to their unique immunological properties CFPs can be expressed in a wide range of prokaryotic (Stahl and Murray, 1989; Francis and Clarke, 1989) and eukaryotic systems (Newton et al, 1987; Beesley et al, 1990). They also provide the opportunity for N-terminal (Newton et al, 1987; Clarke et al, 1987) and C-terminal (Stahl and Murray, 1989) fusions as well as internal insertions into surface antigenic loops on the particle (Schrodel et al, 1990; Clarke et al, 1991) (Fig. 6). The object of such internal insertions is firstly to restrict the inserted epitope in a more defined conformation and secondly to replace an important immunogenic site on the HBcAg with a "foreign" sequence in order to direct and enhance the immune response against that sequence.

In initial experiments a sequence from human rhinovirus (HRV) VP2 156-170 known to elicit anti-virus neutralizing antibodies (Francis et al, 1987b) was inserted into the e1-loop region of core and its immunogenicity in guinea pigs was compared to that of N-terminal CFP's (Fig. 7). The e1-insert CFP preparation was not only 10-fold more immunogenic but it also appeared that a greater proportion of the anti-HRV peptide antibodies produced reacted with virus. This observation was supported by improved neutralizing activity detected in the antisera (Clarke et al, 1991). Therefore, by modifying the presentation system used we were able both to enhance immunogenicity and direct the immune response against a more structured form of the inserted "foreign" peptide.

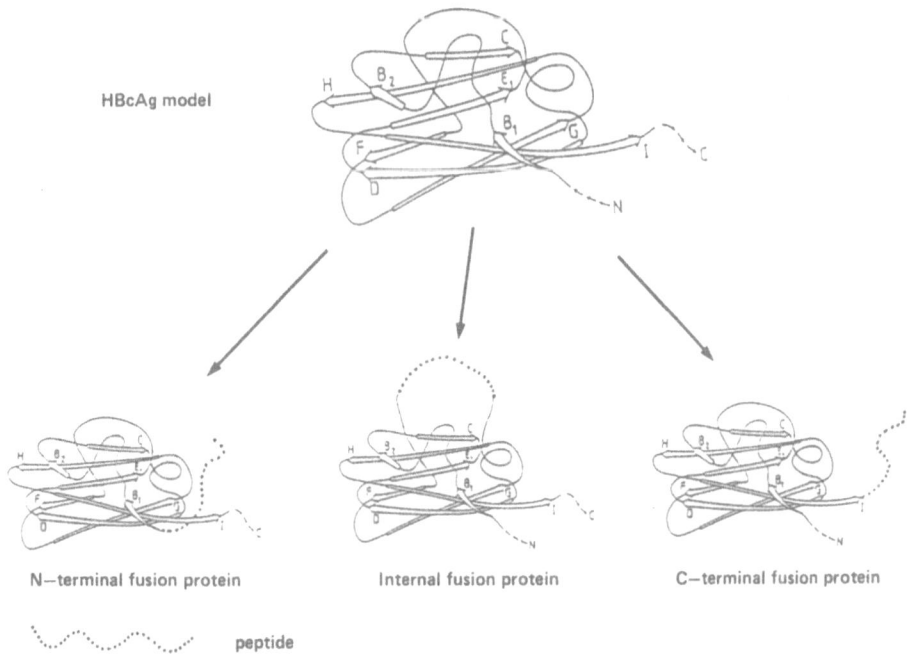

Fig. 6. N-terminal, C-terminal and internal e1-loop fusions of peptide to hepatitis B core antigen (HBcAg).

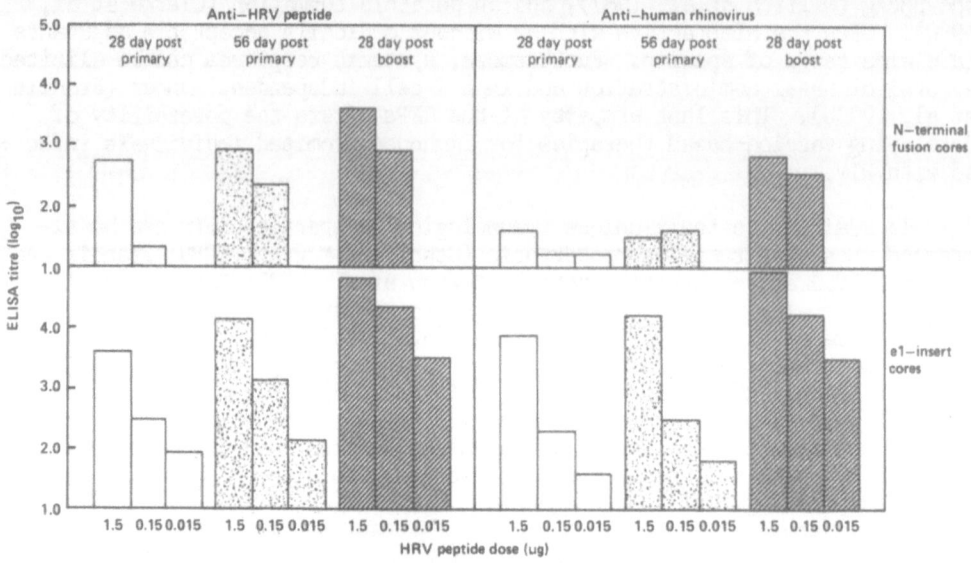

Fig. 7. Anti-HRV peptide and virus responses of guinea pigs to recombinant HBcAg particles (CFPs) with N-terminal fused or e1-inserted HRV2 peptide.

CONCLUSIONS

In this report a range of peptide delivery systems have been compared and contrasted. From the results obtained it is clear that significant quantitative and qualitative differences in the immune response may be produced.

For wholly synthetic vaccines it has been shown that peptides must contain B-cell and Th-cell epitopes (Francis and Clarke, 1989). However, it also appears that multimeric presentation can offer a significant improvement in immunogenicity. The future for such vaccines appears to lie in the ability to construct more complex structures synthetically with defined multiple B-cell epitopes, a suitable range of T-cell epitopes, an adjuvant/immunostimulating sequence and possibly a targeting sequence to modify the immune response, for example, by directing peptides to mucosal surfaces for local responses or directing them to specific areas of the spleen/lymph nodes for enhanced activity.

In taking steps towards synthetic vaccines we were able to learn much from our experiences with defined fusion proteins produced using recombinant DNA technology. The advantages of having multiple B-cell epitopes were clearly demonstrated using the β-galactosidase fusion proteins. This concept plus those of particulate presentation, structural epitopes and Th-cell activity were exploited to great effect by the CFP system. The immunogenicity of peptides could be improved by at least three orders of magnitude and a higher proportion of anti-peptide antibodies that recognise virus could be produced. It seems likely therefore that such presentation systems are likely to provide the first generation of commercially viable peptide vaccines, probably for a veterinary application. Nevertheless, with the clear advantages of a fully synthetic product and the encouraging results obtained to date (see above), future research in this area should eventually lead to the production of safe, effective and chemically defined peptide vaccines for a variety of infectious diseases and other immunotherapeutic applications.

REFERENCES

Adams, S.E., Dawson, K.M., Gull, K., Kingsman, S. and Kingsman, A.J., 1987, The expression of hybrid HIV: Ty virus like particles in yeast, Nature (Lond), 329:68.

Beesley, K.M., Francis, M.J., Clarke, B.E., Beesley, J.E., Dopping-Hepenstal, P.J.C., Clare, J.J., Brown, F. and Romanos, M.A., 1990, Expression in yeast of amino-terminal peptide fusions to hepatitis B core antigen and their immunological properties, Biotechnol., 8:644.

Bittle, J.L., Houghten, R.A., Alexander, H., Schinnick, T.M., Sutcliff, J.G., Lerner, R.A., Rowlands, D.J. and Brown, F., 1982, Protection against foot and mouth disease by immunization with a chemically synthesized peptide predicted from the viral nucleotide sequence, Nature (Lond.), 298:30.

Bittle, J.L., Worrell, P., Houghten, R.A., Lerner, R.A., Rowlands, D.J. and Brown, F., 1984, Immunization against foot-and-mouth disease with a chemically synthesized peptide, in: "Modern Approaches to Vaccines", R.M. Chanock and R.A. Lerner, eds., Cold Spring Harbor Lab., N.Y.

Broekhuijsen, M.P., Blom, T., Kottenhagen, M., Pouwels, P.H., Meloen, R.H., Barteling, S.J. and Engel-Valk, B.E., 1986a, Synthesis of fusion proteins containing antigenic determinants of foot-and-mouth disease virus, Vaccine, 4:119.

Broekhuijsen, M.P., Blom, T., Van Rijn, J., Pouwels, P.H., Klasen, E.A., Fasbender, M.J. and Enger-Valk, B.E., 1986b, Synthesis of fusion proteins with multiple copies of antigenic determinant of foot-and-mouth disease virus, Gene, 49:189.

Broekhuijsen, M.P., Van Rijn, J.M.M., Blom, A.J.M., Pouwels, P.H., Enger-Valk, B.E., Brown, F. and Francis, M.J., 1987, Fusion proteins with multiple copies of the major antigenic determinant of foot-and-mouth disease virus project both the natural host and laboratory animals, J.gen.Virol., 68:3137.

Clarke, B.E., Newton, S.E., Carroll, A.R., Francis, M.J., Appleyard, G., Syred, A.D., Highfield, P.E., Rowlands, D.J. and Brown, F., 1987, Improved immunogenicity of a peptide epitope after fusion to hepatitis B core protein, Nature (Lond.), 330:381.

Clarke, B.E., Brown, A.L., Grace, K.K., Hastings, G.Z., Brown, F., Rowlands, D.J. and Francis, M.J., 1990, Presentation and immunogenicity of viral epitopes on the surface of hybrid hepatitis B virus core particles produced in bacteria, J.gen Virol., 71:1109.

Clarke, B.E., Carroll, A.R., Brown, A.L., Yon, J., Parry, N.R., Rud, E.W., Francis, M.J. and Rowlands, D.J., 1991, Expression and immunological analysis of hepatitis B core fusion particles carrying internal heterologous sequences, in: "Vaccines 91", Cold Spring Harbor Laboratory, New York (in press).

Delpeyroux, F., Chenciner, N., Lim, A., Malpiece, Y., Blondel, B., Grainic, R., van der Wef, S. and Streeck, R.E., 1986, A polio neutralization epitope expressed on hybrid hepatitis B surface antigen particles, Science, 233:472.

DiMarchi, R., Brooke, G., Gale, C., Crocknell, V., Doel, T. and Mowat, N., 1986, Protection of cattle against foot-and-mouth disease by a synthetic peptide, Science, 232:639.

Francis, M.J., 1990, Peptide vaccines for viral diseases, Sci.Progress, Oxford, 74:115.

Francis, M.J. and Clarke, B.E., 1989, Peptide vaccines based on enhanced immunogenicity of peptide epitopes presented with T-cell determinants or hepatitis B core protein, Meth.Enzymol., 178:659.

Francis, M.J., Fry, C.M., Rowlands, D.J., Brown, F., Bittle, J.L., Houghten, R.A. and Lerner, R.A., 1985, Immunological priming with synthetic peptides of foot-and-mouth disease virus, J.gen.Virol., 66:2347.

Francis, M.J., Fry, C.M., Rowlands, D.J., Bittle, J.L., Houghten, R.A., Lerner, R.A. and Brown, F., 1987a, Immune response to uncoupled peptides of foot-and-mouth disease virus, Immunol., 61:1.

Francis, M.J., Hastings, G.Z., Sangar, D.V., Clark, R.P., Syred, A., Clarke, B.E., Rowlands, D.J. and Brown, F., 1987b, A synthetic peptide which elicits neutralizing antibody against human rhinovirus type 2, J.gen.Virol., 68:2687.

Francis, M.J., Fry, C.M., Rowlands, D.J. and Brown, F., 1988, Qualitative and quantitative differences in the immune response to foot-and-mouth disease virus antigens and synthetic peptide, J.gen.Virol., 69:2483.

Francis, M.J., Hastings, G.Z., Clarke, B.E., Brown, A.L., Beddell, C.R., Rowlands, D.J. and Brown, F., 1990a, Neutralizing antibodies to all seven serotypes of foot-and-mouth disease virus elicited by synthetic peptides, Immunol., 69:171.

Francis, M.J., Hastings, G.Z., Brown, A.L., Grace, K.G., Rowlands, D.J., Brown, F. and Clarke, B.E., 1990b, Immunological properties of hepatitis B core antigen fusion proteins, Proc.Natl.Acad.Sci.USA, 87:2545.

Francis, M.J., Hastings, G.Z., Brown, F., McDermed, J., Lu, Y.A. and Tam, J.P., 1991, Immunological evaluation of the multiple antigen peptide (MAP) system using a major immunogenic site of foot-and-mouth disease virus, J.Immunol, submitted.

Kleid, D.G., Dowbenko, D.J., Bock, L.A., Hotlin, M.E., Jackson, M.L., Patzer, E.J., Shine, S.J., Weddell, G.N., Yansura, D.G., Morgan, D.O., McKercher, P.D. and Moore D.M., 1985, Production of recombinant vaccines from microorganisms: Vaccine for foot-and-mouth disease, in: "Microbiology", L. Leive, ed., Washington, D.C., American Society for Microbiology.

Krzych, U.A., Fowler, A.V., Miller, A. and Sercarz, E.E., 1982, Repertoires of T cells directed against a large protein antigen, ß-galactosidase, J.Immunol., 128:1529.

Manca, F., Kunki, A., Fenoglio, D., Fowler, A., Sercarz, E. and Celada, F., 1985, Constraints in T-B cooperation related to epitope topology on E. coli ß-galactosidase, Eur.J.Immunol., 15:345.

Millich, D.R., McLachlan, A., Moriarty, A. and Thornton, G.B., 1987, Immune response to hepatitis B virus core antigen (HBcAg): Localization of T-cell recognition sites within HBcAg/HBeAg, J.Immunol., 139:1223.

Newton, S.E., Clarke, B.E., Appleyard, G., Francis, M.J., Carroll, A.R., Rowlands, D.J., Skehel, J. and Brown, F., 1987, Novel antigen presentation via vaccinia, in: "Vaccines 87: Modern Approaches to New Vaccines", R.M. Chanock, R.A. Lerner, F. Brown and H. Ginsberg, eds., Cold Spring Harbor Laboratory, New York.

Pfaff, F., Mussgay, M., Bohm, H.O., Schulz, G.E. and Schaller, H., 1982, Antibodies against a preselected peptide recognize and neutralize foot-and-mouth disease virus, EMBO J., 1:869.

Posnett, D.N., McGrath, H. and Tam, J.P., 1988, a novel method for producing anti-peptide antibodies: Production of site specific antibodies to the T cell antigen receptor ß-chain, J.Biol.Chem., 263:1719.

Schrodel, F., Weiner, T., Will, H. and Millich, D., 1990, Recombinant HBV core particles carrying immunodominant B-cell epitopes of the HBV pre-S2 region, in: "Vaccines 90: Modern Approaches to New Vaccines Including Prevention of AIDS", F. Brown, R.M. Chanock, H.S. Ginsberg and R.A. Lerner, eds., Cold Spring Harbor Laboratory, New York.

Stahl, S. and Murray, K., 1989, Immunogenicity of peptide fusions to hepatitis B virus core antigen, Proc.Natl.Acad.Sci.USA, 68:6283.

Tam, J.P., 1988, Synthetic peptide vaccine design: Synthesis and properties of a high-density multiple antigenic peptide system, Proc.Natl.Acad.Sci.USA, 85:5409.

Tam, J.P., 1989, Multiple antigen peptide system: A novel design for peptide-based antibody and vaccine, in: "Vaccines 90: Modern Approaches to New Vaccines Including Prevention of AIDS", R.A. Lerner, H. Ginsberg, R.M. Chanock and F. Brown, eds., Cold Spring Harbor Laboratory, New York.

Winther, M.D., Allen, G., Bomford, R.H. and Brown, F., 1986, Bacterially
    expressed antigenic peptide from foot-and-mouth disease virus capsid
    elicits variable immunologic responses in animals, J.Immunol.,
    136:1835.

.

IMMUNOMODULATION BY ADJUVANTS

R. Bomford, M. Stapleton and S. Winsor

Wellcome Biotechnology
Langley Court
Beckenham, Kent BR3 3BS, UK

INTRODUCTION

The immune response has evolved different mechanisms of protection against intracellular and extracellular pathogens. The latter are controlled by antibody, which neutralises bacterial toxins and viruses and opsonises bacteria to facilitate their engulfment by phagocytes. Intracellular pathogens can also be attacked by antibody, but this must be of an isotype that can mediate the killing of infected cells by complement-dependent lysis or by antibody-dependent cellular cytotoxicity (ADCC). In addition, cell-mediated immunity (CMI) can contribute to protection against intracellular pathogens. This can take the form of CD4-positive T lymphocytes which secrete interferon gamma (IFN-gamma) which activates macrophages to destroy pathogens resident within them, such as mycobacteria or leishmania; or CD8-positive cytotoxic T lymphocytes which directly lyse infected cells.

The vaccines against human diseases which are currently available exert their protective effect largely or exclusively by circulating neutralising antibody, with the sole exception of the BCG vaccine against tuberculosis. However, many of the new vaccines that are now contemplated against viral or parasitic diseases will need to engage other mechanisms of defence. An understanding of the factors regulating the expression of the different components of the immune response is therefore essential for the rational design of vaccines. This chapter examines the role of adjuvants in this process. It first reviews the experimental data which relate to the differential induction of antibody isotypes and CMI by adjuvants, and then discusses the mechanisms by which adjuvants might control the nature of the immune response in the light of what is now known of the regulatory role of cytokines and their secretion by different subsets of T lymphocytes.

IMMUNOMODULATION BY ADJUVANTS

There is now a considerable body of evidence that adjuvants can influence the nature of the immune response in terms of the expression of CMI and antibody isotypes.

In the guinea pig, Freund's incomplete adjuvant (FIA), a water-in-oil emulsion of mineral oil, stimulates a strong antibody response that is confined to IgG1, but the inclusion of mycobacteria in the oil (Freund's

Vaccines, Edited by G. Gregoriadis *et al.*
Plenum Press, New York, 1991

complete adjuvant, FCA) leads to the appearance of IgG2 antibody and delayed-type hypersensitivity (DTH) as well (Benacerraf et al, 1963; White et al, 1963). The modern equivalent of FCA is the SAF-1 adjuvant (see chapter by Byars in this volume), consisting of an oil-in-water emulsion with pluronic 121 and the mycobacteria replaced by a derivative of MDP. This also induces DTH in the guinea pig (Byars and Allison, 1987) and IgG2a antibodies in the mouse (Kenney et al, 1989).

Saponin and its superior formulation ISCOMS is another adjuvant which elicits IgG2 antibody and DTH. In a comparison of a number of adjuvants for the vaccination of mice with a cell-surface glycoprotein from the South American protozoal parasite Trypanosoma cruzi, the causative agent of Chaga's disease, saponin stim-ulated IgG2a and DTH and was the best adjuvant for protective immunity (Scott et al, 1984). The protection was correlated with DTH because squalane induced equally high IgG2a titres, but no DTH or protection. ISCOMS have been shown to stimulate IgG2a antibodies to viral antigens in the mouse (Lovgren, 1988). In a recent study on the effect of adjuvants on the isotype of the antibody response of mice to fluorescein isothiocyanate (FITC)-labelled human gamma globulin (HGG) or dextran sulphate (DXS), it was found that the predominant plaque-forming cells (PFC) were IgG1 for FIA, FCA, Al(OH)$_3$ and saponin (Quil A), although Al(OH)$_3$, FCA and Quil A did induce some IgG2a PFC (Karagouni and Hadjipetrou-Kourounakis, 1990). Very striking results were seen with polynucleotide adjuvants, poly I:C causing a response that was almost exclusively IgG2a, whereas poly A:U (and BeSO$_4$) favoured IgG2b. LPS and LiCl directed the response towards IgG3. These results are remarkable for the degree of control exerted by the adjuvants over the isotype of the antibody response, and also for the fact that this control was identical whether the antigen was thymus dependent (FITC-HGG) or thymus independent (FITC-DXS).

Our own work on the relative efficacy of adjuvants for the gp120 antigen of HIV in mice and the isotypes of the antibody response has revealed that in BALB/c mice saponin and an oil-in-water emulsion with pluronic 121 plus MDP were both effective (Fig. 1) but that saponin stimulated both IgG1 and IgG2a, and the emulsion system only IgG1 (Fig. 2).

The IgE response is regulated by adjuvants in the mouse, Al(OH)$_3$ promoting this isotype whereas FCA does not (Hamaoka et al, 1974; Kishimoto et al, 1976). The only circumstances under which it might be desirable deliberately to stimulate IgE antibody by vaccination would be by a vaccine against schistosome worms, when IgE may play a role in protective immunity. In one study in the mouse with a Schistosoma mansoni vaccine, it was found that Al(OH)$_3$, but not FCA, generated IgE anti-schistosome antibodies and protection (Horowitz et al, 1982).

THE REGULATION OF ANTIBODY ISOTYPE BY CYTOKINES

The role of cytokines in the regulation of the antibody isotype secreted by B lymphocytes has been studied by adding cytokines to in vitro cultures of B lymphocytes undergoing a polyclonal mitogenic response to LPS. Interleukin 4 (IL-4) and interleukin 5 (IL-5) promoted the production of IgG1 and IgE (Snapper et al, 1988) whereas IFN-gamma stimulated IgG2a (Snapper and Paul, 1987). These are all cytokines produced by T lymphocytes and when establshed murine T cell clones were examined for cytokine secretion, it was found that they could be divided into two subsets: Th1 cells secreting IL-4 and IL-5, and Th2 cells IFN-gamma and interleukin 2 (IL-2) (Bottomley, 1988; Mosmann and Coffman, 1989). In addition, the Th1 cells mediated DTH (Cher and Mosmann, 1987).

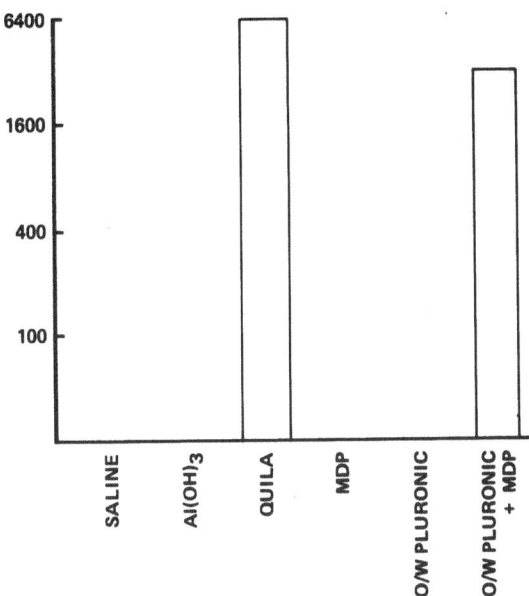

Fig. 1. The secondary immune response of BALB/c mice to HIV gp120
in various adjuvants. Mice were primed with 0.5 μg of gp120
(CHO recombinant) in saline, Al(OH)$_3$ (100 μg Al), saponin
(Quil A, 50 μg), MDP (Sigma, 50 μg), a squalane-in-water
emulsion containing pluronic 121 (Byars and Allison, 1987)
or a mixture of MDP and emulsion. Twenty eight days later,
they were boosted with 1 μg gp120 with the same adjuvants
and bled 14 days later. The results shown are ELISA titres
of pooled sera of groups of five mice.

There is a striking parallelism between the functions of the Th1 and
Th2 subsets and the selective induction of components of the immune response
by adjuvants. Al(OH)$_3$ stimulates IgG1 and IgE which are controlled by the
Th2 subset. FCA and saponin also stimulate an IgG1 response, but in addition they elicit IgG2a and DTH, which are promoted by the Th1 subset. It may
be hypothesised that Al(OH)$_3$ favours the triggering of Th2 cells, whereas
FCA and saponin engage Th1 cells as well (Table 1). There is now some

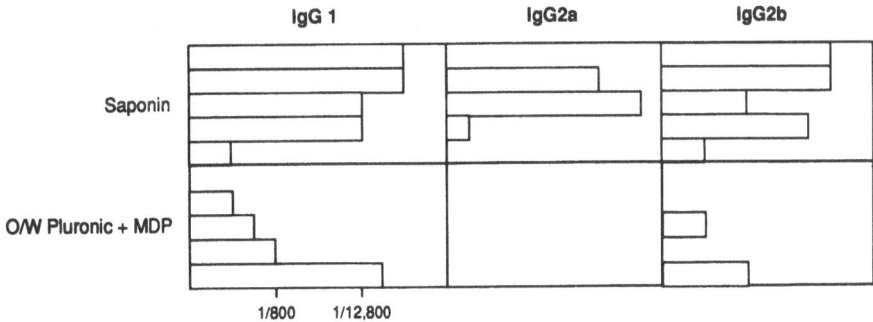

Fig. 2. The isotype distribution of the secondary immune response to
gp120 adjuvanted with saponin or MDP emulsion in BALB/c mice.
Experimental details as in Fig. 1. The results are ELISA
titres of individual mice.

Table 1. Summary of the Induction of Antibody Isotypes and Delayed-type Hypersensitivity (DTH) by Different Adjuvants in Mice and Guinea Pigs and the Possible Control by T Cell Subsets

Adjuvant	Induction of Antibody Isotypes and Delayed Type Hypersensitivity				T Cell Subsets Involved?	
	IgG1	IgE	IgG2a	DTH	Th1	Th2
Al(OH)$_2$	+	+	-	-	+	-
FIA	+	-	-	-	+	-
FCA	+	-	+	+	+	+
SAF-1	+	-	+	+	+	+
Saponin (ISCOMS)	+	-	+	+	+	+

experimental support for this hypothesis. The antigen-specific proliferative response of lymph node cells of mice immunised with Al(OH)$_3$ could be inhibited by antibodies against IL-1, in contrast to the cells from FCA-immunised mice, which were unaffected (Grun and Maurer, 1989). Since Th2 cells respond to IL-1 as a growth factor and Th1 cells do not, these results are consistent with the induction of Th2 cells by Al(OH)$_3$ and Th1 cells by FCA.

An alternative hypothesis that might apply in some cases of selective induction of isotypes by adjuvants is that the adjuvants are bypassing the T cell compartment by inducing other cells to induce cytokines which then cause isotype switching by B cells. Thus, it is difficult to acocunt for the control of the isotype response to thymus-independent antigens such as FITC-DXS (Karagouni and Hadjipetrou-Kourounakis, 1990) by selective induction of antigen-specific T cell subsets but, for instance, the promotion of the IgG2a response to this antigen by poly I:poly C could be the result of IFN-gamma secretion by NK cells.

The two T cell subsets can interact in a mutually antagonistic manner through the cytokines they produce. IL-4 secreted by Th2 cells blocks the effect of Th1-derived IFN-gamma on IgG2a production by B cells (Snapper and Paul, 1987) and a recently discovered Th1 cytokine, named cytokine secretion inhibition factor (CSIF), or now IL-10, inhibits the production of all cytokines by Th2 cells (Fiorentino et al, 1989). This functional cross-inhibition between the two Th subsets may explain the classical observation that the immune response often becomes canalised into antibody production or CMI (Parish, 1972). There are also examples of the canalisation of the immune response by adjuvants. For instance, immunisation of guinea pigs with FIA which does not induce DTH prevents the appearance of DTH after subsequent immunisation with FCA, a phenomenon which was named "immune deviation" (Asherson and Stone, 1965). Similarly, in the mouse, prior immunisation with the antigen associated with mycobacteria blocks the induction of IgE by subsequent immunisation with Al(OH)$_3$ (Kishimoto et al, 1976). However, it is obvious that, although IgE and DTH are suppressible, IgG1 and IgG2a antibodies can co-exist which implies at least a partial functional co-expression of Th1 and Th2 cells.

Although the theory of the control of antibody isotypes by cytokines released by separate T cell subsets was developed on the basis of in vitro experiments, there is now increasing evidence for the existence of Th1 and Th2 cells in vivo.

First, chemical modification of the antigen can change the nature of the immune response. Manoeuvres which increase the hydrophobicity of the antigen such as acylation or acetoacetylation reduce antibody formation and increase DTH (Parish, 1971; Coon and Hunter, 1973), which suggests that the response is being biased towards Th1 cells. Secondly, parasitic infection provides some interesting examples of selective T cell triggering. Leishmania infection in mice is controlled by IFN-gamma-producing Th1 cells, and IL-4-producing Th2 cells exacerbate the disease (Liew, 1989). BALB/c mice, which are relatively susceptible to infection, are protected by a vaccine administered intravenously, which preferentially stimulated Th1 cells; but subcutaenous immunisation causes increased susceptibility to infection. In this case, it is the route of injection which controls the predominating T cell subset. In contrast to Leishmania, the immune response to the nematode worm Nippostrongylus braziliensis is characterised by a strong IgE response, which is mediated by Th2 cells (Finkelman et al, 1990). Finally, the histocompatibility background can regulate the induction of T cell subsets. Immunisation of I-A^b mice with collagen type IV leads to an in vitro lymphocyte proliferative response of the Th1 type (IFN-gamma-secreting) and no antibody; in I-A^b mice, there is no lymphocyte proliferative response, although the lymphocytes secrete IL-4, and IgG1 anbtibody is produced (Murray et al, 1989).

What hypotheses have been advanced to account for selective induction of T cell subsets and how can these accommodate the effect of adjuvants? First, it is possible that selection is exercised by the specificity of the antigen, i.e. there are some peptide epitopes that can only be recognised by Th1 cells and others only by Th2. This implies a difference in the receptor repertoires of the two T cell subsets. If true, this hypothesis would have important consequences for vaccinology, because it would open the way to adjusting vaccines to stimulate the appropriate response by including the correct epitopes. It easily accounts for the control of T cell subsets by histocompatibility backgrounds, because this in turn controls which epitopes can bind to class II MHC antigens and be presented to T cells. As for adjuvants, it could be proposed that they create different peptides by differential processing, although it is difficult to see how this effect could be consistent for Th1 or Th2 epitopes for different antigens. Also, in general it appears unlikely that evolution should have imposed a difference in repertoires in the T cells responsible for the control of intracellular and extracellular pathogens. Altogether, although the repertoire hypothesis cannot be formally excluded until the requisite studies on the stimulation of T cells by defined peptide epitopes have been accomplished, it does seem to present considerable difficulties.

The alternative is to ensure that the two T cell subsets respond to different differentiative or triggering signals supplied by antigen-presenting cells (APC). This could involve a different type of APC for the two subsets (Bottomley and Janeway, 1989) or the modulation of the function of a single type of APC by the antigenic stimulus (Weaver et al, 1988). The available evidence is consistent with the idea that Th2 cells, which respond to IL-1, while Th1 cells are independent of IL-1 but require a non-IL-1 costimulatory signal from the APC.

There is already some evidence that adjuvants can modulate the function of APC in a manner that might favour the induction of Th1 and Th2 cells.

First, a number of adjuvants such as LPS, MDP and $Al(OH)_3$ have been shown to stimulate the synthesis of IL-1 by monocytes (Mannhalter et al, 1985; Bahr et al, 1987) and this could stimulate Th2 cells and account for the stimulation of the IgG1 response by most adjuvants. Secondly, in studies on the induction of contact sensitivity, which is likely to be mediated by Th1 cells, it was found that peritoneal cells induced by FCA, when labelled with trinitrophenyl (TNP), were much more potent at inducing the reaction (Britz et al, 1982).

CONCLUSIONS

Adjuvants are one of a number of factors which regulate antibody isotype and CMI in rodents and in some cases, if not all, the control may be exerted by the selective induction of T cell subsets. Can these ideas be extrapolated to man? The question as to whether adjuvants can regulate antibody isotypes and CMI in man must await the clinical application of new adjuvants (see chapter by Byars for the SAF-1 adjuvant). It remains controversial as to whether human T cells can be functionally divided into two subsets as those of rodents. Human T cell clones have been established whose cytokine secretion pattern does not conform to Th1 or Th2 (Paliard et al, 1988). However, recent work comparing lymphokine secretion by T cell clones from atopic or normal individuals has yielded IL-4 or IFN-gamma-secreting clones respectively (Plaut, 1990).

REFERENCES

Asherson, G.L. and Stone, S.H., 1965, Selective and specific inhibition of 24-hour skin reactions in the guinea pig. I. Immune deviation: description of the phenomenon and the effect of splenectomy, Immunology, 9:205.

Bahr, G.M., Chedid, L.A. and Behbenani, K., 1987, Induction, in vivo and in vitro, of macrophage membrane interleukin-1 by adjuvant-active synthetic muramyl peptides, Cell.Immunol., 107:443.

Benacerraf, B., Ovary, Z., Bloch, K.J. and Franklin, E.C., 1963, Properties of guinea pig 7S antibodies. I. Electrophoretic separation of two types of guinea pig 7S antibodies, J.exp.Med., 117:937.

Bottomley, K., 1988, A functional dichotomy in $CD4^+$ T lymphocytes, Immunol.Today, 9:268.

Bottomley, K. and Janeway, C.A., 1989, Antigen presentation by B cells, Nature, 337:24.

Britz, J.S., Askenase, P.W., Ptak, W., Steinman, R.M. and Gershon, R.K., 1982, Specialized antigen-presenting cells. Splenic dendritic cells and peritoneal-exudate cells induced by mycobacteria activate effector T cells that are resistant to suppression, J.exp.Med., 155:1344.

Byars, N.E. and Allison, A.C., 1987, Adjuvant formulation for use in vaccines to elicit both cell-mediated and humoral immunity, Vaccine, 5:223.

Cher, D.J. and Mosmann, T.R., 1987, Two types of murine helper T cell clones. II. Delayed type hypersensitivity is mediated by Th1 clones, J.Immunol., 138:3688.

Coon, J. and Hunter, R., 1973, Selective induction of delayed hypersensitivity by a lipid conjugated protein antigen which is localised in thymus dependent lymphoid tissue, J.Immunol., 110:183.

Finkelman, F.D., Holmes, J., Katona, I.D., Urban, J.F., Beckman, M.P., Schooley, K.A., Coffman, R.L., Mosmann, T.R. and Paul, W.E., 1990, Lymphokine control of in vivo immunoglobulin selection, Ann.Rev.Immunol., 8:303.

Fiorentino, D.F., Bond, M.W. and Mosmann, T.R., 1989, Two types of mouse T
    helper cell. IV. Th2 clones secrete a factor that inhibits cytokine
    production by Th1 clones, J.exp.Med., 170:2081.
Grun, J.L. and Maurer, P.H., 1989, Different T helper cell subsets elicited
    in mice utilising two different adjuvant vehicles: The role of
    endogenous interleukin 1 and proliferative responses, Cell.Immunol.,
    121:134.
Hamaoka, T., Newburger, P.E., Katz, D.H. and Benacerraf, B., 1974, Hapten-
    specific IgE antibody responses in mice. III. Establishment of
    parameters for generation of helper T cell function regulating the
    primary and secondary responses of IgE and IgG B lymphocytes,
    J.Immunol., 113:958.
Horowitz, S., Smolarsky, M. and Arnon, R., 1982, Protection against
    Schistosoma mansoni achieved by immunisation with sonicated parasite,
    Eur.J.Immunol., 12:327.
Karagouni, E.E. and Hadjipetrou-Kourounakis, L., 1990, Regulation of isotype
    immunoglobulin production by adjuvants in vivo, Scand.J.Immunol.,
    31:745.
Kenney, J.S., Hughes, B.W., Masada, M.P. and Allison, A.C., 1989, Influence
    of adjuvants on the quantity, affinity, isotype and epitope speci-
    ficity of murine antibodies, J.Immunol.Methods, 121:157.
Kishimoto, T., Hirai, Y., Suemara, M. and Yamamaura, Y., 1976, Regulation of
    antibody response in different immunoglobulin classes. I. Selective
    suppression of anti-DNP IgE antibody response by preadministration of
    DMP-coupled mycobacterium, J.Immunol., 117:396.
Liew, F.Y., 1989, Functional heterogeneity of CD4[+] T cells in leishmaniasis,
    Immunol.Today, 10:40.
Lovgren, K., 1988, The serum antibody response distributed in subclasses and
    isotypes after intranasal and subcutaenous immunisation with
    influenza virus immunostimulating complexes, Scand.J.Immunol.,
    27:241.
Mannhalter, J.W., Neychev, H.O., Zlabinger, G.J., Ahmad, R. and Eibl, M.M.,
    1985, Modulation of the human immune response by the non-toxic and
    non-pyrogenic adjuvant aluminium hydroxide: Effect on antigen uptake
    and antigen presentation, Clin.exp.Immunol., 61:143.
Mosmann, T.R. and Coffman, R.L., 1989, Heterogeneity of cytokine secretion
    patterns and functions of helper T cells, Adv.Immunol., 46:111.
Murray, J.S., Madri, J., Tite, J., Carding, S.R. and Bottomley, K., 1989,
    MHC control of CD4[+] T cell subset activation, J.exp.Med., 170:2135.
Paliard, X., de Waal Malefijt, R., Yssel, H., Blanchard, D., Chretien, I.,
    Abrams, J., de Vries, J. and Spits, H., 1988, Simultaneous production
    of IL-2, IL-4 and IFN-gamma by activated human CD4[+] and CD8[+] T cell
    clones, J.Immunol., 141:849.
Parish, C.R., 1971, Immune response to chemically modified flagellin. II.
    Evidence for a fundamental relationship between humoral and cell-
    mediated immunity, J.exp.Med., 134:21.
Parish, C.R., 1972, The relationship between humoral and cell-mediated
    immunity, Transplant.Rev., 13:35.
Plaut, M., 1990, Antigen-specific lymphokine secretory patterns in atopic
    disease, J.Immunol., 144:4497.
Scott, M.T., Bahr, G., Moddaber, F., Afchain, D. and Chedid, L., 1984,
    Adjuvant requirements for protective immunisation of mice using a
    Trypanosoma cruzi 90K cell surface glycoprotein, Int.Arch.Allergy
    appl.Immunol., 74:373.
Snapper, C.M., Finkelman, F.D. and Paul, W.E., 1988, Differential regulation
    of IgG1 and IgE synthesis by interleukin 4, J.exp.Med., 167:183.
Snapper, C.M. and Paul, W.E., 1987, Interferon-gamma and B cell stimulatory
    factor-1 reciprocally regulate Ig isotype production, Science,
    236:944.

Weaver, C.T., Hawrylowicz, C.M. and Unanue, E.R., 1988, T helper cell sub-
    sets require the expression of distinct costimulatory signals by
    antigen-presenting cells, Proc.Natl.Acad.Sci.USA, 85:8181.
White, R.G., Jenkins, G.C. and Wilkinson, P.C., 1963, The production of
    skin-sensitizing antibody in the guinea-pig, Int.Arch.Allergy,
    22:156.

USE OF SYNTEX ADJUVANT FORMULATION TO ENHANCE IMMUNE RESPONSES TO VIRAL

ANTIGENS

N.E. Byars, G.M. Nakano, M. Welch and A.C. Allison

Syntex Research
Palo Alto
California, USA

INTRODUCTION

There is both the need and the potential for the development of new
vaccines and the improvement of existing vaccines. One aspect of this
process is the development of better adjuvant formulations so that both
humoral and cell-mediated immunity are enhanced. We have been interested in
the adjuvant formulation area of vaccine development for some time, and
believe that the Syntex Adjuvant Formulation (SAF) has a number of advantages that will make it very useful for both human and veterinary vaccines.
SAF has been shown to increase both antibody synthesis and cell-mediated
immunity to a variety of antigens in a number of species (Byars and Allison,
1987, Allison and Byars, 1988, Morgan et al, 1989). Cell-mediated responses
have been measured by delayed hypersensitivity skin tests in vivo and by in
vitro blastogenesis and/or interleukin-2 synthesis in response to antigen
challenge. Animals have also been protected against live virus challenge,
or against challenge with tumor cells, in situations where cell-mediated
responses are a necessary component of protective immunity (Byars and
McRoberts, 1986; Morgan et al, 1989; Campbell et al, 1990). SAF also
increases efficacy of vaccines, compared to alum or saline control vaccines,
thus allowing a reduction in either the amount of antigen per dose or in the
number of doses required. The reduction in amount of antigen required is
particularly advantageous if the antigen is scarce or is difficult or expensive to purify. A reduction in cost and in the number of doses needed may
be especially important in developing countries where health care expendit-
ures are very limited. Other advantages of SAF include the lack of toxicity
observed so far, and in the ease of use of the emulsion, which has the consistency of milk. The vaccine is prepared by gently mixing the antigen with
the pre-made emulsion containing the appropriate threonyl muramyl dipeptide
(Thr-MDP) concentration. Thus, the antigen is not denatured. This is in
marked contrast to Freund's-type emulsions, where the antigen is present
during the emulsification process and may be partially or completely denatured during vaccine preparation (Kenney et al, 1989). Furthermore, the
SAF emulsion is very stable. It can be frozen, refrigerated, left at room
temperature, or even 40°C, without affecting the physical integrity of the
emulsion. Optimal stability of the Thr-MDP is achieved at 4°C and/or a pH
between 5 and 7. Of course, the stability of the antigen will largely
dictate the storage time and temperature for any given vaccine. The process
of preparing SAF emulsions has progressed from laboratory scale methods

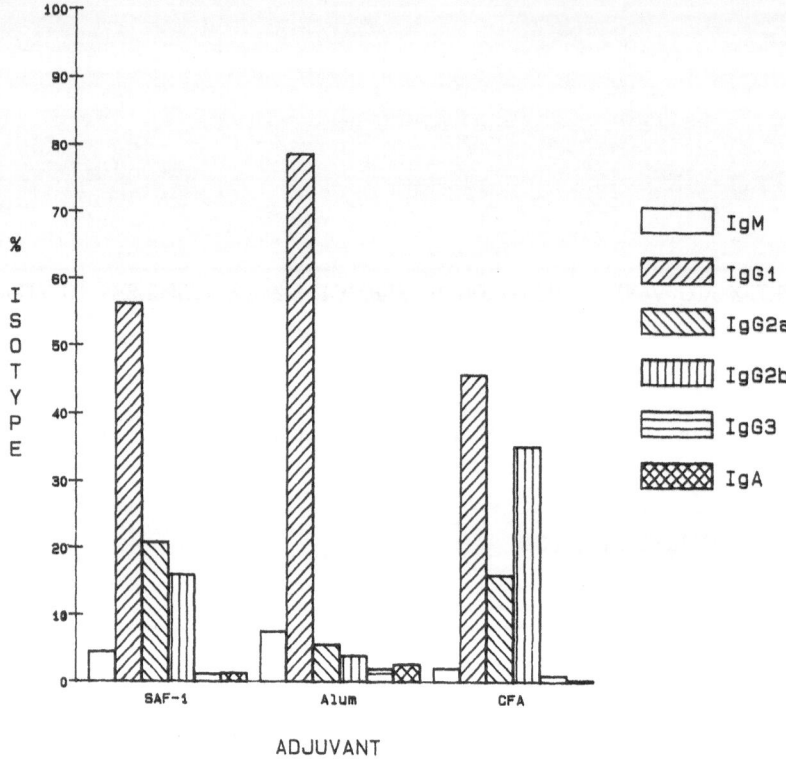

Fig. 1. Isotypes of anti-HBsAg antibodies in BALB/c mice following
immunization with HBsAg in SAF, alum or CFA.

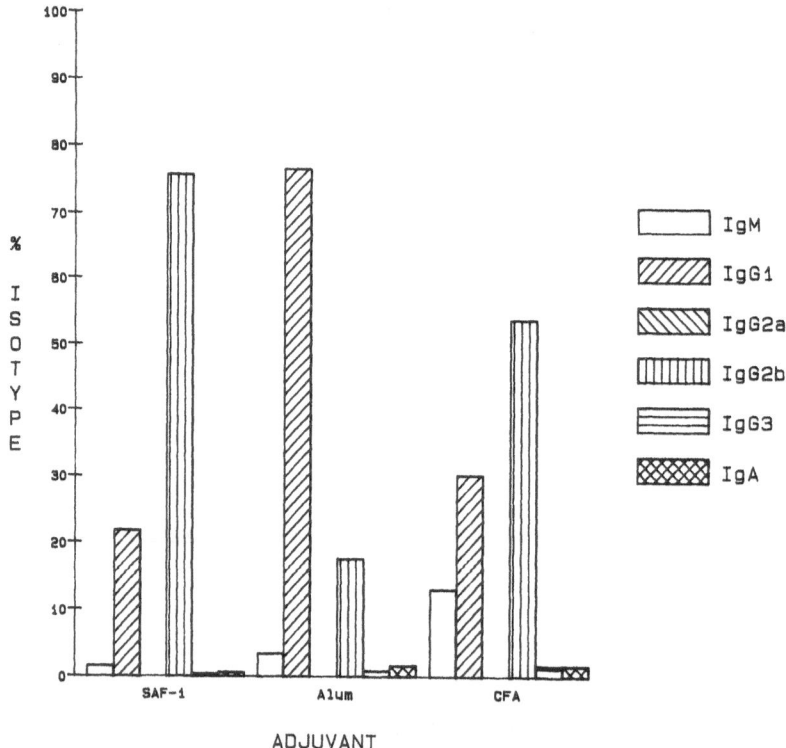

Fig. 2. Isotypes of anti-HBsAg antibodies in B10.M mice following
immunization with HBsAg in SAF, alum or CFA.

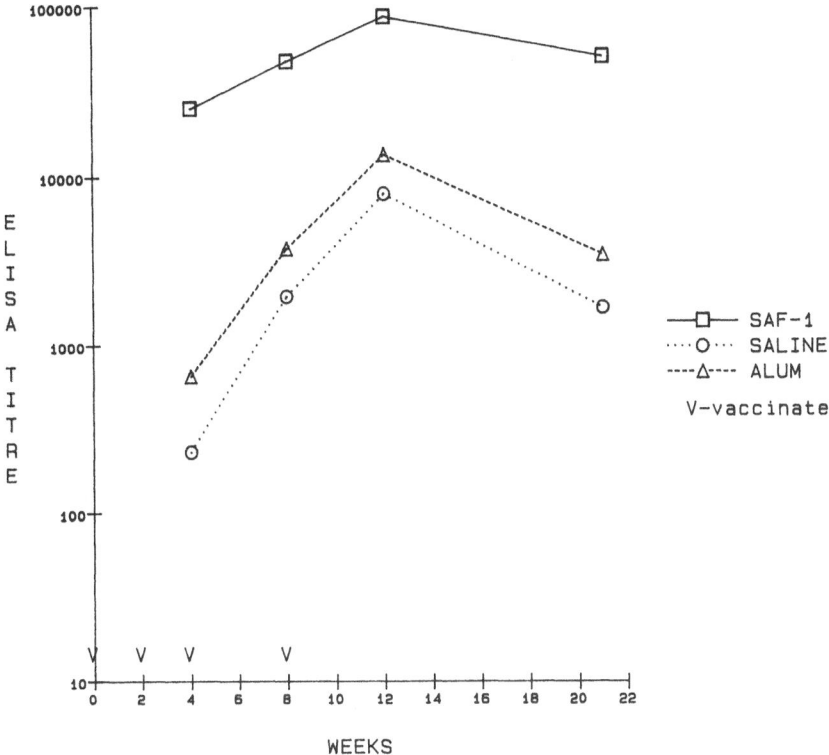

Fig. 3. Anti-gag titres of pooled guinea pig sera, following immunization of groups of 8 female guinea pig with the gag protein
in SAF, alum or saline. The arrows indicate the times at which
the animals were given primary or booster immunizations.

(principally, vigorous vortexing of the components) to methods suitable for
scale-up to commercial production quantities (Lidgate et al, 1989).

One aspect of immunization which has been of interest to us is how to
improve the responses of "low responders". Low responders include young
children, people over 65 years of age, and those who are genetically predisposed to make poor responses to particular antigens. As experimental
models, we have used young mice (three weeks old) and old mice (13 months
old) to examine responses to influenza hemagglutinins and hepatitis B virus
surface antigen (HBsAg). The B10.M/SnH^{2f} strain of mice were reported to be
non-responders to HBsAg in CFA (Milich and Chisari, 1982) so these mice were
selected as a model of genetic non-responsiveness to examine responses to
HBsAg in SAF or alum.

Influenza vaccines presently available for human use are simply saline
solutions of the viral hemagglutinins (HA) or suspensions of killed viruses.
We found that less than half of the young mice given 1.0 or 0.05 µg of
influenza B HA in saline produced measurable anti-HA titres. In contrast,
all the young mice given 1.0 or 0.05 µg of HA in SAF responded with anti-HA
titres (Byars et al, 1990a). Influenza can be the cause of significant
morbidity and mortality in elderly people, and studies have found that
vaccine efficacy in the elderly ranges from 25% to 70%. When we compared
influenza HA in SAF or saline in old BALB/c mice, we found that all the mice
(11/11) given 1 µg HA in SAF produced anti-HA antibodies, while few (2/10)
of the mice given 1 µg HA in saline did so. Further, the mice given the HA
in SAF had significantly higher titres than those given HA in saline (Byars
et al, 1990a). The same pattern of consistent response was seen in young

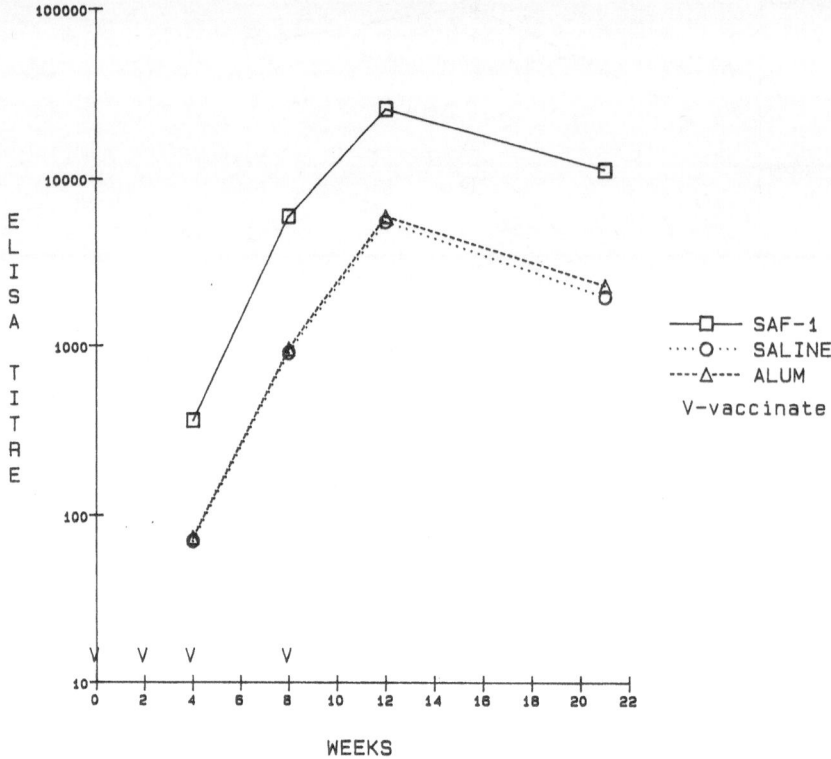

Fig. 4. Anti-env titres of pooled guinea pig sera, following immunization of groups of 8 female guinea pigs with the env protein
in SAF, alum or saline. The arrows indicate the times at which
the animals were given primary or booster immunizations.

BALB/c mice given 1 μg of HBsAg in SAF, compared to less uniform responses
and lower titres observed in the young mice given HBsAg in alum (Byars et
al, 1990b).

The issue of genetic non-responsiveness was addressed by testing HBsAg
vaccines in the B10.M strain of mice. We found they responded quite well to
HBsAg in SAF and less well to HBsAg on alum. The responses were lower than
those of BALB/c mice, but were readily measured. With the B10.M mice, we
found that s.c. vaccination was much more efficacius than i.p. injections,
and that the females generally responded better than did the males. Thus,
the adjuvant, the route of administration and the sex of the mice affect
non-responsiveness.

Some immunoglobulin isotypes are more likely to provide protection
against infections and tumors than others. In the mouse, the desirable
isotype is IgG2a, while in the human, it is IgG1. It would be ideal to be
able to control the isotype of specific antibodies by the selective use of
adjuvants in vaccines. We examined the isotype of anti-HBsAg antibodies
produced when mice were vaccinated with HBsAg in SAF, alum or complete
Freund's adjuvant (CFA). The data we obtained indicate that judicious
selection of the adjuvant can have a profound effect on the isotype of specific antibodies induced by a vaccine. In BALB/c mice, considerably more
IgG2a antibodies against HBsAg were induced with SAF compared to those
induced using alum. Alum induced predominantly IgG1 antibodies, while CFA
induced more IgG1 and IgG2b than IgG2a (see Fig. 1). Relatively little
anti-HBsAg of the IgG1 isotype was detected in sera of B10.M mice vaccinated
using SAF, while significant amounts of antigen-specific IgG1 were found.

In contrast, sera from mice vaccinated using alum had a great deal of antigen-specific IgG1 and relatively little IgG2b. The isotype profile of mice vaccinated using CFA was intermediate between the SAF and alum profiles (Fig. 2). We were unable to detect IgG2a in the sera of B10.M mice. This may be because the polyclonal anti-IgG2a sera used react strongly with IgG2a of the Igh^a allotype and weakly with IgG2a of the Igh^b allotype. B10.M mice are of the Igh^b allotype while BALB/c mice are of the Igh^a allotype. Experiments to quantitate the IgG2a^b of the B10.M mice are in progress, using allotype-specific antisera.

To determine whether SAF or alum would be a more effective adjuvant for antigens derived from HIV, we immunized guinea pigs with the "gag" and "env" proteins, synthesized by recombinant DNA technology. Control animals were given the proteins in saline. Animals were vaccinated with 50 μg of protein at week 0 and with 25 μg of protein at 2, 4 and 8 weeks, then bled at 4, 8, 12 and 21 weeks. For both proteins, SAF proved to be the better adjuvant. The difference was greater for gag proteins, with titres being 6.5 to 38-fold higher at various time points when SAF was the adjuvant than when alum was used (Fig. 3). With the env protein, SAF induced 5 to 6-fold higher antibody titres than alum (Fig. 4). For the env protein, alum proved to be ineffective as an adjuvant, since the titres were essentially the same as those obtained when env was given in saline. We also vaccinated four cynomolgus monkeys with the gag protein. Three animals received 100 μg of gag in SAF on days 0, 14, 28 and 42, while one received the protein in saline containing Thr-MDP. All three monkeys given the gag/SAF vaccine produced anti-gag antibodies, while the one monkey given gag in saline produced no detectable antibodies, even after four doses (Fig. 5).

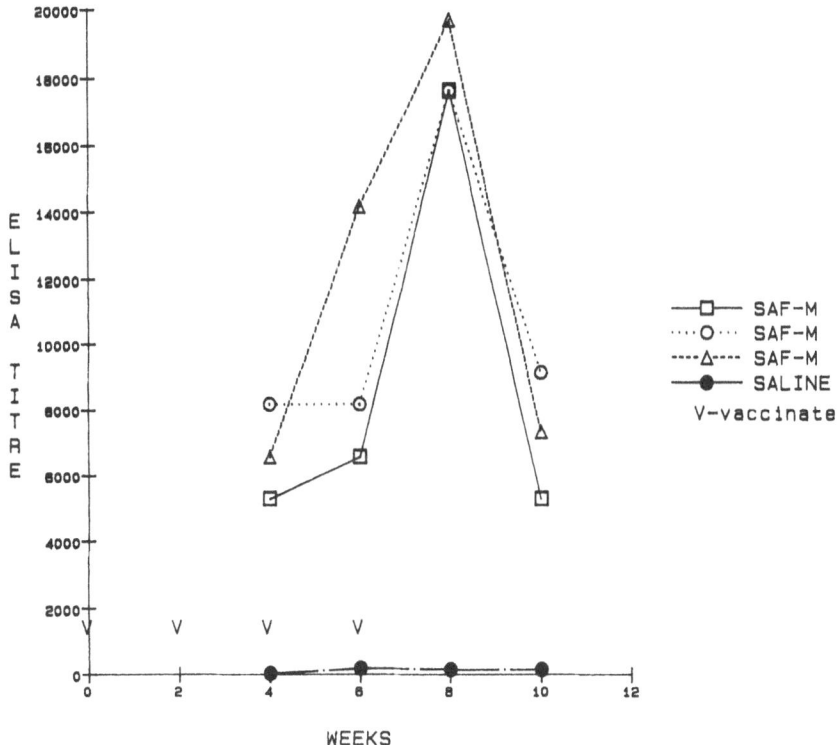

Fig. 5. Anti-gag titres of individual cynomolgus monkeys, immunized with gag in SAF or in saline. The arrows indicate the times at which the animals were given primary or booster immunizations.

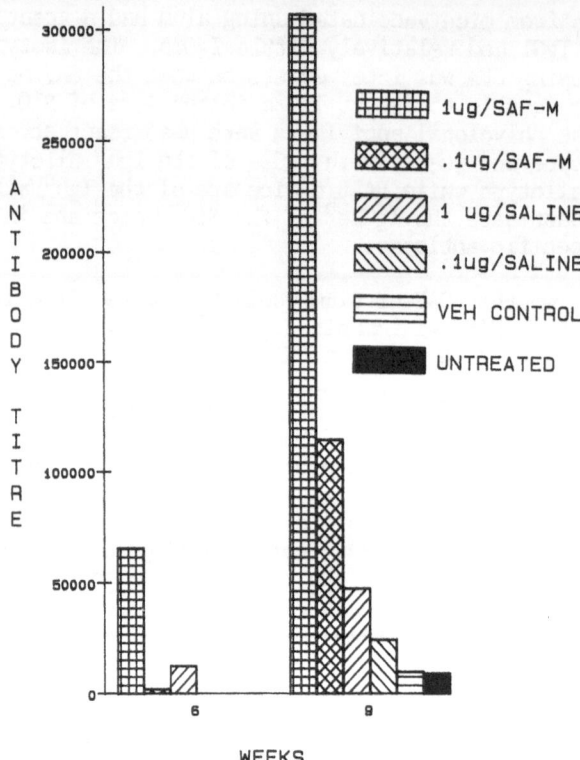

Fig. 6. Anti-gDt titres of pooled sera of groups of 20 female guinea
pigs immunized with gD-2t in SAF or saline. Control groups
were given SAF alone or left untreated.

Because herpes simplex virus, type 2 (HSV-2) has become an increasingly
significant problem among sexually transmitted diseases, a vaccine against
HSV-2 would be valuable. We have tested a vaccine containing the gD-2t
protein of HSV-2 in SAF in guinea pigs and measured antibody titres, resistance to challenge with live virus, and in vitro lymphocyte proliferation in
response to gD-2t. This truncated, recombinant protein was very kindly
provided by Dr. Rae Lyn Burke of Chiron Corporation. The guinea pigs were
vaccinated at 0 and 4 weeks, bled at 6 weeks, one day prior to vaginal
challenge with live virus, then at 9 weeks, i.e. 3 weeks post-challenge.

We found that with either 0.1 or 1.0 µg of gD-2t in SAF, significant
antibody titres were induced by week 6, prior to live virus challenge.
These titres were 3 or 10-fold higher than those obtained using 0.1 or 1.0
µg gD-2t in saline. At week 9 (3 weeks post-challenge), the groups immunized with the SAF-adjuvanted vaccines still had 5 to 6-fold higher titres
than the saline groups, despite the booster effect of live virus challenge.
Control groups developed anti-gD-2t titres following challenge with the live
virus. These titres were somewhat lower than those of the groups given gD-
2t in saline (Fig. 6).

Lymphocyte proliferation in vitro in response to concanavalin A (Con A)
or various concentrations of gD-2t was used to measure cell-mediated responses to the antigen. Cells were obtained from four randomly chosen
animals from each of the untreated and vehicle-only control groups and the
groups given 0.1 µg gD-2t in SAF or saline. Cells from all the control
animals responded to Con A, but not to gD-2t (Fig. 7). Cells from the
animals given 0.1 µg gD-2t in saline responded strongly to Con A but only
slightly to gD-2t in vitro. Cells from two of the four animals vaccinated

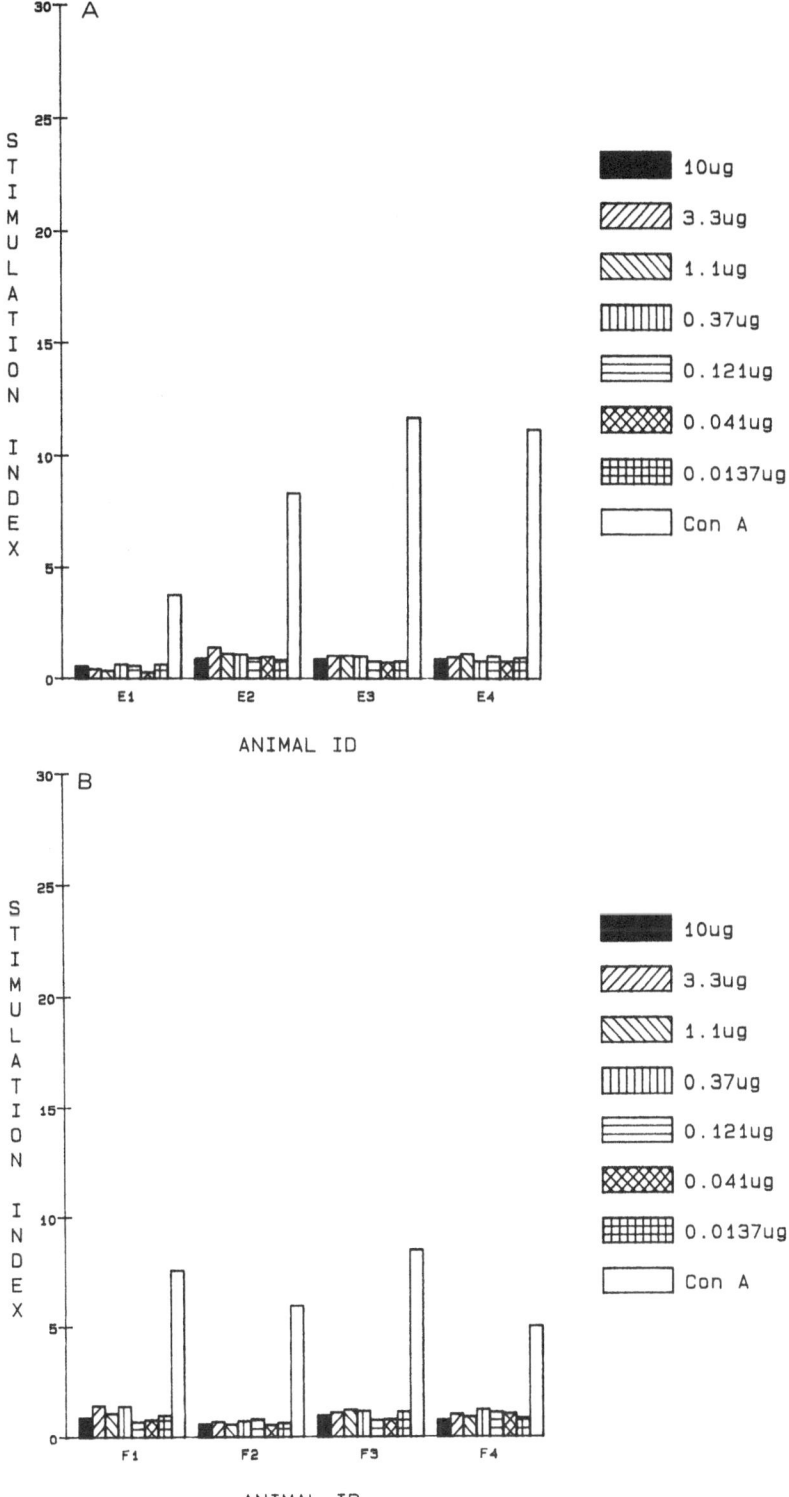

Fig. 7. (a) Proliferation of leukocytes from 4 guinea pigs given SAF only. Cells were obtained one day prior to live virus challenge. (b) Proliferation of leukocytes from 4 untreated guinea pigs, one day prior to live virus challenge.

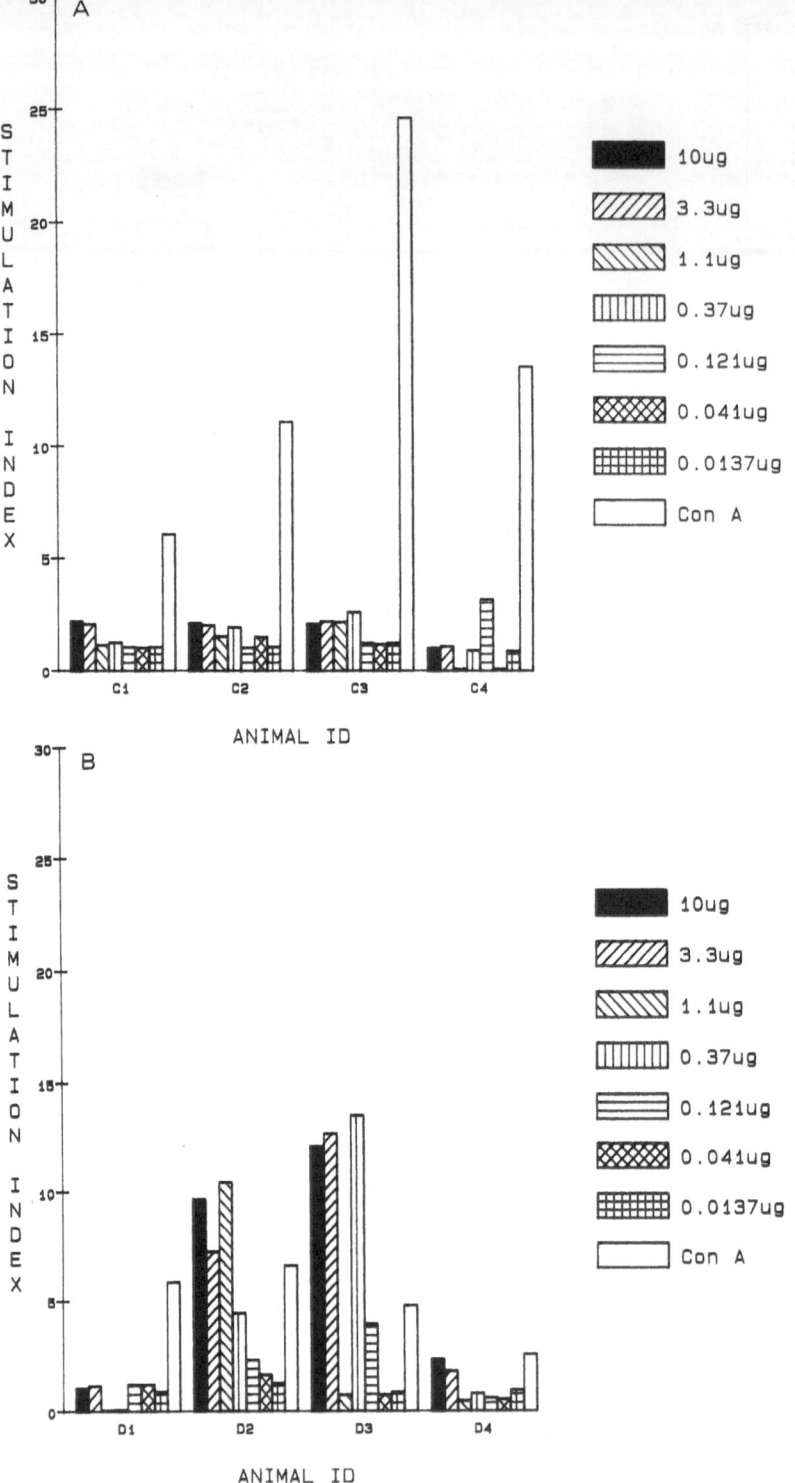

Fig. 8. (a) Proliferation of leukocytes from 4 guinea pigs vaccinated with 0.1 μg gD-2t in saline. Cells were obtained one day prior to live virus challenge.

(b) Proliferation of leukocytes from 4 guinea pigs vaccinated with 0.1 μg gD-2t in SAF. Cells were obtained one day prior to live virus challenge.

with 0.1 μg gD-2t in SAF had strong responses to gD-2t in vitro and all four responded to Con A (Fig. 8). Three weeks post-challenge, cells were obtained from another 4 randomly chosen animals from each group. Again, animals from the two control groups had a response to Con A, but essentially no response to gD-2t, despite having antibody titres against gD-2t as a result of infection with live virus. Two of four of the animals from the gD-2t/saline group had antigen-specific responses to in vitro antigen challenge. However, all four animals from the gD-2t/SAF group had antigen-specific responses, which were greater than those of the saline group.

The pre-challenge antibody titres and the in vitro cell proliferation data correlate well with the vaginal lesion scores. The lesion scores showed that the groups vaccinated using SAF were better protected against live virus challenge than those vaccinated with gD-2t in saline. Fewer animals in the SAF groups had lesions, and those lesions that did occur were much milder than those observed in animals in the gD-2t/saline or control groups.

These data suggest that while antibodies to HSV are necessary, they are not sufficient for protection against disease. For complete protection, both humoral and cell mediated immunity are necessary. It is also interesting that the two control groups developed anti-gD-2t antibodies in the 22 days following challenge with live virus, but did not develop CMI as measured by cellular proliferation. This lack of CMI may be a factor in the development of latency following natural infections with HSV. We found that HSV could be cultured from the dorsal root ganglia of most of the control animals (17/22) but rarely from the ganglia of animals vaccinated using SAF (1/20).

The data presented above indicate that SAF is a very efficacious adjuvant for viral antigens, for induction of both humoral and cell mediated responses. The lack of toxicity observed so far in animal studies, the stability of the emulsion, and the ease with which vaccines can be prepared without denaturing the antigens are all significant advantages of SAF. Further, the ability to induce antibodies of a desired isotype is also a useful property of SAF. If the findings in the animals can be reproduced in humans, then protection against a number of viral diseases will be possible.

REFERENCES

Allison, A.C. and Byars, N.E., 1988, An adjuvant formulation for use with subunit and recombinant antigens, in: "Technological Advances in Vaccine Development", L. Lasky, ed., Alan R. Liss, New York.
Byars, N.E. and Allison, A.C., 1987, Adjuvant formulation for use in vaccines to elicit both cell mediated and humoral immunity, Vaccine, 5:223.
Byars, N.E., Allison, A.C., Harmon, M.W. and Kendal, A.B., 1990a, Enhancement of antibody responses to Influenza B virus hemagglutinin by use of a new adjuvant formulation, Vaccine, 8:49.
Byars, N.E. and McRoberts, M.J., 1986, Cell mediated responses to a feline leukemia virus vaccine, Abstracts of the Annual Meeting of the American Society for Microbiology, 99.
Byars, N.E., Nakano, G., Welch, M., Lehman, D. and Allison, A.C., 1990b, Improvement of Hepatitis B vaccine by the use of a new adjuvant, Vaccine, in press.
Campbell, M.J., Esserman, L., Byars, N.E., Allison, A.C. and Levy, R., 1990, Idiotype vaccination against murine B cell lymphoma. Humoral and cellular requirements for the full expression of antitumor immunity, J.Immunol., 145:1029.

Kenney, J.S., Hughes, B.W., Masada, M.P. and Allison, A.C., 1989, Influence of adjuvants on the quantity, affinity, 150 type and epitope specificity of murine antibodies, J.Immunol.Meth., 121:157.

Lidgate, D.M., Fu, R.C., Byars, N.E., Foster, L.C. and Fleitman, J.S., 1989, Formulation of vaccine adjuvant muramyldipeptides. 3. Processing optimization, characterization, and bioactivity of an emulsion vehicle, Pharm.Res., 6:748.

Milich, D.R. and Chisari, F.V., 1982, Genetic regulation of the immune response to hepatitis B surface antigen (HBsAg). 1. H-2 restriction of the murine humoral response to the a and d determinants of HBsAg, J. Immunol., 129:320.

Morgan, A.J., Allison, A.C., Finerty, S., Scullion, F.T., Byars, N.E. and Epstein, M.A., 1989, Validation of a first generation Epstein-Barr virus vaccine preparation suitable for human use, J.Med.Virol., 29:74.

INFLUENZA VACCINES AND THE WYETH-AYERST EXPERIENCE WITH SYNTEX ADJUVANT

Richard N. Hjorth, Geraldine M. Bonde, Elizabeth D. Piner,
Kenneth M. Goldberg, Daniel M. Teller, Mark Hite, Steven K.
Vernon, Mark H. Levner and Paul P. Hung

Wyeth-Ayerst Research
P.O. Box 8299
Philadelphia, PA 19101, USA

BACKGROUND

Importance of the Disease

Influenza/pneumonia is the only infectious disease(s) among the top ten
causes of death in the U.S. (Check, 1984). Influenza virus kills an average
of over 20,000 people each year in the United States (Katz, 1985). In
addition, the illness has a substantial economic impact. Schoenbaum (1987)
estimated that the total direct cost of influenza (health care cost) exceeds
$1 billion per year in the U.S., with the actual cost (including lost pro-
ductivity) on the order of $3 to $5 billion per year. Yet, until recently
at least, only about 20% of the over 43 million Americans at risk of death
from influenza by virtue of their age or underlying medical condition
receive vaccinations (Mostow, 1986; Check, 1984). If there is a close match
between the vaccine strain and the circulating strain, protection can be 70
to 80% in the adult population (Douglas, 1990). Protection may be lower in
the infirm elderly who are at greatest risk (Gross et al, 1989).

Reasons for Lack of Vaccine Utilization

There are many reasons why influenza vaccine is one of the least
effectively utilized vaccines (Katz, 1985). Public fear of side effects
persists despite the major improvements in influenza vaccines which occurred
19 or more years ago. Double-blinded studies in the young and elderly show
that the only significant side effect with modern vaccines is a sore arm
(Margolis et al, 1990; Parkman et al, 1976). Guillain-Barre' syndrome is
still remembered as being associated with swine flu vaccination in 1976,
although this association has been called into question (Kurland et al,
1985) and no association with other influenza vaccines has been seen (Kaplan
et al, 1982). Perceived lack of efficacy is a second major reason for vaccine under-utilization. This perception in part reflects the fact that the
vaccine is not fully efficacious but also stems from the public misconcep-
tion that influenza vaccines should prevent all winter respiratory illnesses
(Katz, 1985). Other reasons include the fact that people are not familiar
with the possible fatal disease course in high risk patients, while many
physicians themselves feel that vaccination is not worth the effort. Also,
Medicare does not reimburse for vaccination.

Vaccines, Edited by G. Gregoriadis *et al.*
Plenum Press, New York, 1991

To rectify the problem of vaccine under-utilization, many groups including the Center for Disease Control sparked by A.P. Kendal, the National Foundation for Infectious Diseases, the American Lung Association, the manufacturers and others have recently tried to correct some of the public misconceptions through advertising. Indeed, vaccine utilization seems to be increasing somewhat but an even stronger effort is needed. Lobbying by these groups has resulted in a pilot study on reimbursement by Medicare. The other approach required is to actually improve vaccine efficacy, a goal many groups have worked on for years. This article will describe the current vaccines and discuss some of the methods used to try to improve their efficacy. The Wyeth-Ayerst experience with the Syntex adjuvant will then be described.

Current Influenza Vaccines in the U.S.

There are four manufacturers marketing influenza virus subunit vaccines in the United States: Wyeth-Ayerst, Connaught, Parke-Davis and Lederle. Connaught also markets a killed whole virus vaccine. Whole virus vaccines are generally considered to be more reactogenic in children than subunit vaccines (Miles et al, 1981). All of the subunit vaccines are similar in reactogenicity and efficacy. A major advance occurred in the late 1970's when the FDA began requiring standardization using the single radial immuno-diffusion (SRID) test instead of the more variable chick cell agglutination (CCA) test to measure the quantity of the major protective antigen (hemagglutinin [HA]). Modern vaccines in the U.S. consist of 15 μg of HA per dose of each component. Previously, whole virus vaccines artifactually seemed to be more efficacious than subunit vaccines because the CCA test sometimes resulted in the whole virus vaccines containing more HA. Today, because of the SRID test, whole virus and subunit vaccines have equivalent efficacy (Quinnan et al, 1983; Cate et al, 1983). Three strains are presently used in vaccines representing H_1N_1 (H=HA; N=neuraminidase) and H_3N_2 subtypes of influenza type A and the B type. The vaccines are all manufactured using egg grown virus because the yields are superior to those obtained in tissue culture. Vaccines are purified using various combinations of centrifugation, chromatography, precipitation, filtration and ultrafiltration. Virus for subunit vaccines is disrupted with Tween 80 and ether, Tween 80 and tri-n-butyl-phosphate, or Triton X-100. The resulting product consists of aggregated hemagglutinin and neuraminidase spikes which bind through their hydrophobic regions and form rosettes (Fig. 1).

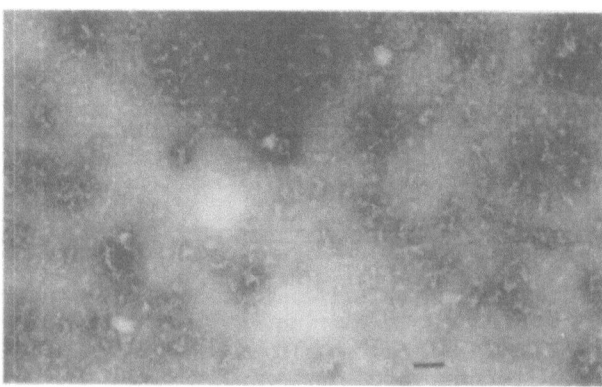

Fig. 1. Electron micrograph of a subunit influenza virus vaccine
         (trivalent).
         Bar represents 100 nm.

Increased Dosage

One of the most obvious means to try improving influenza vaccine efficacy is to increase the dosage. Unfortunately, most attempts to do so have failed because the dose-response curve in humans was relatively flat (Sullivan et al, 1990). When amounts of antigen large enough to cause a significant increase in the antibody response were used, the reactogenicity and the cost of the vaccine became prohibitive (Mostow et al, 1973). For example, Matzkin and Nili (1984) showed that increasing by 10-fold the amount of vaccine injected into young adults resulted in increases in geometric mean hemagglutination inhibition (HI) titers (GMT) of roughly 100% for the H_3N_2 component, 200% for the H_1N_1 component but only 12.5% for the influenza B component. Historically, it has been most difficult to achieve a good antibody response to the B component. Somewhat more success was achieved in seronegative children by Gross et al (1982) who increased the B-component GMT 200% with a 9-fold increase in vaccine antigen.

Increased Number of Doses

Although two doses are recommended for children, the advantage of two doses in adults who have pre-existing antibody is questionable (Levine et al, 1987; Cate et al, 1977; Anon, 1989; Powers et al, 1984).

Different Routes of Vaccination

A considerable amount of effort was expended years ago to develop nasal administration of killed vaccines (Waldman and Coggins, 1972). Nasal administration resulted in the production of secretory antibody which reduced virus replication in the lungs and nasopharynx. The drawbacks to this approach were the large amounts of antigen required to elicit a good response, the reduced effect in unprimed individuals and the fleeting nature of the secretory response. Furthermore, inadequate amounts of serum antibody were induced. In a few small studies some success was achieved using a combination of nasal and intramuscular immunization in the elderly but this approach has not been pursued (Fulk et al, 1969).

Oral administration would probably be the ideal way to deliver vaccine for maximum patient acceptance. Despite the efforts of a number of groups (Walkman et al, 1987; Chen and Quinnan, 1989), this option suffers from the same drawbacks as the intranasal route. Perhaps an adjuvant for the secretory immune system such as cholera toxin subunit B (Hirabayashi et al, 1990), ISCOMS (Lovgren, 1988), immunosomes (Guink et al, 1989) or the use of microencapsulation will one day increase the efficacy of the nasal or oral route enough to be practical.

Liposome Vaccines

Earlier animal studies by our group and others with various types of liposomes did not show convincing improvements when compared with conventional vaccines (Thibodeau et al, 1981). More recently, Pietrobon et al (1989) seem to have developed a promising multilamellar liposome system for influenza vaccines.

Adjuvants

There has not been a new adjuvant approved for general use by the FDA of the U.S.A. in over 40 years. Freund's incomplete adjuvant (FIA) and a water-in-mineral-oil emulsion have been effective with influenza vaccine in clinical trials (Salk and Salk, 1977; Hobson, 1973) but concerns over the

Table 1. Syntex Adjuvant: Composition

Threonyl-Muramyl Dipeptide (t-MDP): an MDP analog with Reduced Toxicity

5.0% Squalane	Vehicle: binds antigen to
2.5% Pluronic L121	microspheres which are
0.2% Tween 80	presented to the immune system

remote possibility of their carcinogenicity and the long persistence of these non-metabolizable oils have prevented licensure in the U.S.A. (Murray et al, 1972). Mineral oil adjuvants were used commercially in Britain between 1963-65 and a low incidence of delayed nodule reactions was reported (Stuart-Harris, 1969). Hilleman's group (1972) obtained good results with "Adjuvant 65" which contained readily metabolized peanut oil, but the possible carcinogenicity of the Arlacel A component apparently prevented licensure.

Alum and aluminum phosphate are the only two adjuvants approved for use in humans in the U.S. today. Unfortunately, alum seems to have no positive effect on the efficacy of influenza vaccines in humans (Jennings et al, 1981). In fact Bachmayer (1976) showed that for his CTAB-extracted subunit vaccine, the adjuvant significantly suppressed the immune response of humans. In contrast, the response by mice to this vaccine was enhanced by alum. This study illustrates species-to-species differences in the response to adjuvants.

More recently, a number of muramyl dipeptide (MDP) analogs with low toxicity have been used in adjuvants such as the Syntex adjuvant, which is effective with influenza virus vaccines (Byars et al, 1990), ISCOMS (Lövgren, 1988), and monophosphoryl lipid A (Masihi et al, 1986). Other adjuvants have also been tested with influenza vaccines.

Live Virus Vaccines

Over the years, a number of live virus vaccines have been investigated (Stuart-Harris, 1980). These include egg-passage attenuated (Salk et al, 1944), inhibitor resistant (Prevost et al, 1973), temperature sensitive (Wright and Karzan, 1987), cold-adapted (Maassab et al, 1990) and avian human recombinants (Sears et al, 1988). The cold-adapted viruses developed by Maassab seem to show the most promise for increasing vaccine efficacy in children. However, in the elderly, pre-existing immunity limits live vaccine effectiveness.

THE WYETH-AYERST EXPERIENCE WITH SYNTEX ADJUVANT

Composition

Many of our studies were carried out with a formulation of Syntex adjuvant (SAF-m) which has the composition shown in Table 1. The function of each component of the adjuvant has been described by Allison and Byars (1986). The amount of threonyl-MDP (t-MDP), or temurtide, used in each mouse was usually 100 µg. The vehicle is thought to function by binding antigen to microspheres which are then presented to the immune system. The adjuvant is microfluidized in order to produce a more stable emulsion. Microfluidization improves the consistency of results compared to SAF-1, an older formulation. Figure 2 shows an electron micrograph of the adjuvant. Unlike Freund's adjuvant, which is thick and difficult to work with, SAF-m

46

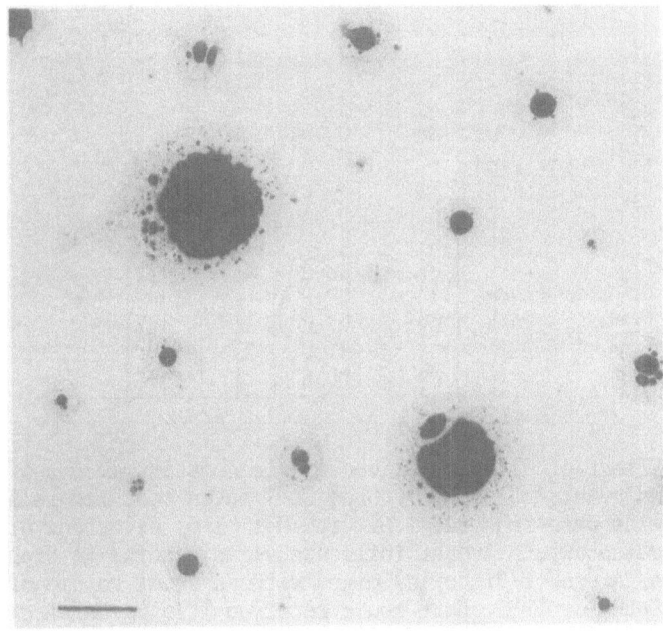

Fig. 2. Electron micrograph of the syntex adjuvant mixed with
influenza virus vaccine.
Bar represents 1000 nm.

is quite fluid and is easily mixed with antigen solutions by a few swirls of
the container.

Efficacy

Most of our studies used the serological mouse potency test to study
the commercial Wyeth-Ayerst trivalent vaccine. The mouse model is perhaps
the most frequently used because of its convenience and general correlation
of results with results seen in seronegative humans (McLaren et al, 1978).
Mice become infected with most unadapted human strains of influenza virus
but they do not become ill. Some strains have been mouse-adapted, denoting
they will kill mice. Little work has been done with these strains because
of the variability in results seen by many investigators. Other animal
models for influenza vaccination include hamsters, guinea pigs and ferrets.

McLaren et al (1978) have shown that there is a good correlation in
mice between hemagglutination inhibition antibody titers and resistance to
challenge infection. In man and in mice, HI titers of 1:40 are generally
considered protective against infections. Our mice were immunized intra-
muscularly rather than intraperitoneally, as was done by McLaren and others,
in order to mimic the human situation and to allow the adjuvant to exhibit
its full effect. We found that the use of the intramuscular route gave
lower serum HI titers than the IP route. In 28 to 35 days individual sera
were assayed for HI antibodies and the results were analyzed statistically.

Figure 3 shows the results of a small experiment in which dilutions of
trivalent vaccine were made in a constant amount of adjuvant. The GMT of HI
antibodies was increased 88-fold at a 1:2 dilution of vaccine. Dilutions of
unadjuvanted vaccine beyond 1:2 have not elicited detectable HI responses.
However, the HI response to A/Sichuan in mice immunized with the adjuvanted

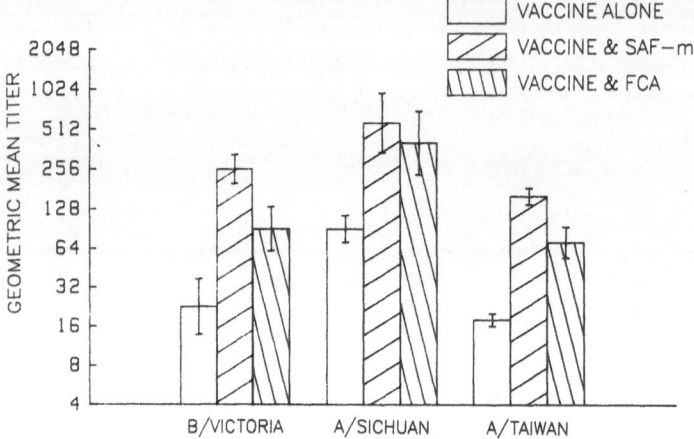

Fig. 3. Effect of dilution of vaccine in constant amount of
adjuvant. Groups of three CD-1 seven-week-old female
mice were injected I.M. with 0.1 ml of dilutions of trivalent Wyeth-Ayerst influenza virus vaccine (A/Sichuan,
A/Taiwan, B/Victoria) mixed with an equal volume of 2X
SAF-m or PBS. Each mouse received 1.75 μg HA/strain at
the 1:2 dilution and 100 μg t-MDP. Thirty-five days
later, serum samples were assayed for antibody to
A/Sichuan by the HI test.

vaccine decreased in a linear fashion as the vaccine was diluted and was
still detectable at a 1:2000 dilution of vaccine. It was estimated that a
1:1500 dilution of vaccine in SAF-m gave an effect equivalent to undiluted
vaccine. Figure 4 demonstrates that the immune response to vaccine in
adjuvant is so strong as to be comparable to, or in the case of the A/Taiwan
component, superior to the response to vaccine in Freund's complete adjuvant
(FCA).

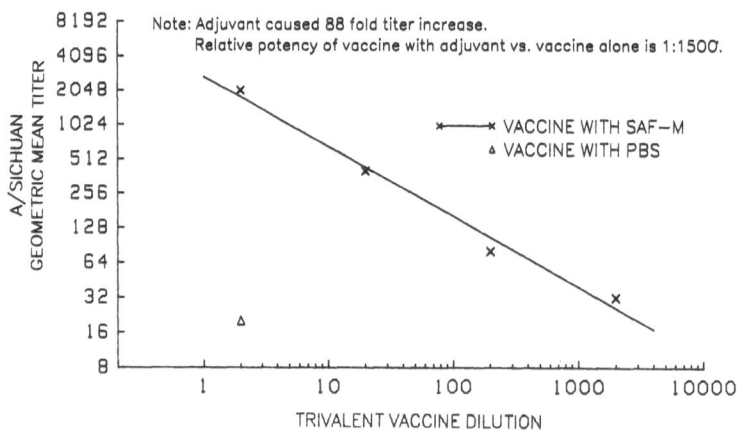

Fig. 4. Comparison of syntex adjuvant with Freund's complete
adjuvant. Groups of six CD-1 seven-week-old female mice
were injected I.M. with 0.2 ml of influenza virus vaccine
mixed with an equal volume of 2X SAF-m, PBS or Freund's
complete adjuvant (Difco). Each mouse received 3.5 μg
HA/strain and 100 μg t-MDP. Twenty-eight days later,
serum samples were assayed for antibody by the HI test.

Table 2. Activity of Components of SAF-m.

	n	A/Taiwan
SAF-m	10	1024
Vehicle	10	776
t-MDP	9	87
Vaccine alone	9	87

Groups of nine to ten CD-1 seven-week-old female mice were
injected I.M. with 0.1 ml of influenza virus vaccine mixed
with an equal volume of 2X SAF-m, 2X vehicle only, t-MDP only,
or PBS. Each mouse received 1.75 μg HA/strain and 100 μg t-MDP.
Twenty-nine days later, serum samples were assayed for HI antibody.
Note. Means connected by a line are not significantly different
at the .05 level. Means not connected are significantly different
at the .001 level.

To determine which components of the adjuvant were active, we conducted
the experiment shown in Table 2. Vaccine plus t-MDP alone had no significant activity. Vaccine with vehicle alone was as active as vaccine with
complete adjuvant. t-MDP has been shown by others to be more active in the
guinea pig. It is unknown whether it will enhance the humoral response in
humans. t-MDP is also important in cellular immune responses and anamnestic
responses, which were not measured here.

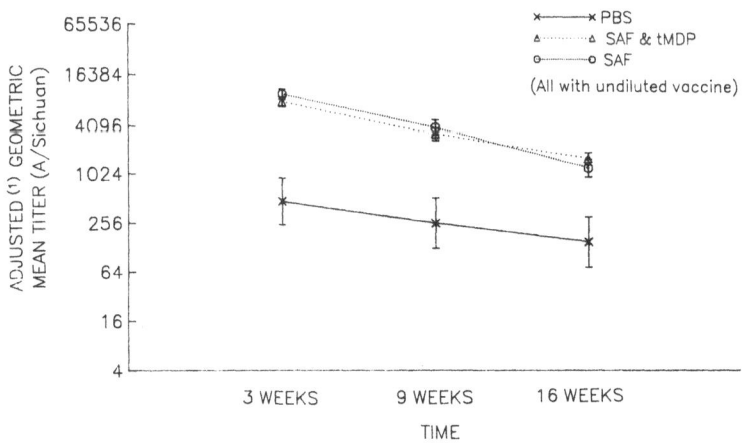

Fig. 5. Stability of antibodies induced with SAF-m. Groups of six to
ten CD-1 seven-week-old female mice were injected I.M. with
0.2 ml of influenza virus vaccine with an equal volume of 2X
SAF-m, 2X vehicle, or PBS. Each mouse received 1.5 μg
HA/strain and 100 μg t-MDP. Mice were eye bled at various
times and the serum frozen until assay in one test. Results
were adjusted statistically for a few missing samples.

[1]Adjusted for mice with missing values.

Table 3. Resistance of Mice to Infection after Vaccination

Group	<1:8 2	<1:8 4	1:8 2	1:8 4	1:16 2	1:16 4	1:32 2	1:32 4	1:64 2	1:64 4	1:128 2	1:128 4	1:256 2	1:256 4
1 (vaccine + SAF-m)											0/4*	0/2		0/2
2 (1:1000 vaccine + SAF-m)			1/2	0/2	1/2	1/2	2/4	0/4	0/1	0/2				
3 (vaccine + PBS)			4/4	1/3	1/5	2/6	0/3	1/2						
4 (PBS + SAF-m)	6/6	5/5												
5 (PBS)	6/6	6/6												
6 (PBS + vehicle)	3/3	2/2												

*number infected/number challenged

Notes: Mice challenged with A/Sichuan 10 weeks after immunization
       No significant difference between groups 2 and 3 by two-tailed Mantel-Haenszel test.

Groups of CD-1 seven-week-old female mice were injected I.M.
with 0.1 ml of undiluted or diluted influenza virus vaccine
mixed with an equal volume of 2X SAF-m or PBS. Each mouse
received 1.75 μg HA/strain and 100 μg t-MDP. Nine weeks later
the mice were eye-bled and the serum assayed for HI antibodies
to A/Sichuan. At week 10 the mice were challenged I.N. with
0.1 ml of A/Sichuan virus (1.2 x 10^7 pfu/ml). Lungs were
removed on day 2 or 4 and assayed for virus in eggs. For
statistical analysis mice were grouped by antibody titer.

Figure 5 indicates that the decay curve of antibodies elicited with or
without adjuvant did not differ statistically, at least for the first 16
weeks after immunization. This result would suggest that the response in
humans should last long enough to protect for at least one influenza season
and possibly longer.

We wanted to confirm our serological mouse potency data with protection
data and to demonstrate the efficiency of antibody elicited with SAF-m. We
injected a number of mice using vaccine with and without adjuvant and
grouped the mice having low titers of antibody (Table 3). If antibody was
going to be less efficient in preventing infection, the effect should be
most noticeable in this borderline region. Mice were challenged with
A/Sichuan virus and the lungs were examined for the presence of virus on
days 2 or 4. As expected, control mice with high titers of antibody were
protected. Mice not immunized with vaccine were not protected. A varying
amount of protection was seen in the mice with borderline amounts of antibody. The two-tailed Mantel-Haenszel statistical test was used to show that
in mice having the same antibody level there was no significant difference
between the group injected with vaccine alone and the group injected with
adjuvanted vaccine. This result demonstrated that there was no difference
in the effectiveness of the antibody elicited with SAF-m.

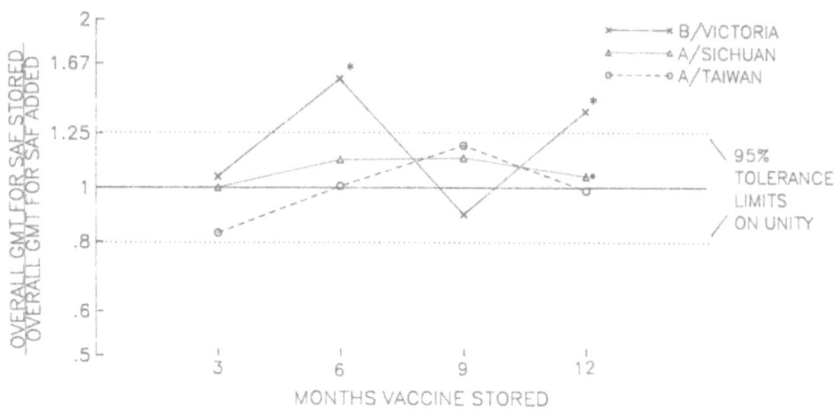

Fig. 6. Influenza vaccine stability study mouse potency tests.
Plot of geometric mean titer ratios comparing SAF stored
with vaccine to SAF added at time of testing. (See legend
below.)
*Titers for SAF stored significantly higher than titers
for SAF added (p<.05) .

Stability

Syntex had conducted studies which showed that the adjuvant alone is
chemically stable for long periods of time, especially when stored at 4°C
under nitrogen. We wanted to examine the biological stability of our
vaccine stored with the adjuvant and, at the same time, the biological
stability of the adjuvant. To do this we prepared a number of doses of
vaccine in the two types of containers in which vaccine is sold by Wyeth-
Ayerst (vials and Tubex). Some of the glass and rubber components of vials
and Tubex differ. A portion of the vaccine was stored with adjuvant and
another portion was stored without adjuvant. Before testing in mice, vaccine was diluted 1:2 in either vehicle or concentrated adjuvant so that the
amount of t-MDP and vehicle was equivalent for both "stored" and "added"
samples. Further dilutions of 1:20 and 1:200 were then made in adjuvant.

Figure 6 illustrates the stability of the vaccine by comparing the
ratio of the overall GMT (all three dilutions) induced by vaccine stored in
adjuvant to vaccine with adjuvant added at the time of testing. It is
obvious that the biological activities of all three components of the 1988-
1989 vaccine were stable when stored with the adjuvant for at least a year.
Conversely, these results also showed that the adjuvant itself remained
biologically stable when stored with vaccine over the one year period.
There was little difference between the data for Tubex and vials. This
shows compatibility of adjuvanted vaccine with the rubber and glass of the
two containers.

Finally, the results from the stored and added groups were pooled to
produce a plot of the GMT at each dilution. Each point of the graph of
A/Taiwan for example (Fig. 7) represents 36 to 72 mice thereby giving us
very small standard errors and highly significant information. We could not
compare the results from month to month as we did in Figure 6 using ratios
because of differences in the batches of mice, stress levels in mice, red
blood cells used in the HI test, etc. However, the results in Fig. 7
confirm the strong effect of the adjuvant on the HI titer. The average
increase in titer for all time points at the 1:2 dilution of vaccine was 15-
fold for A/Taiwan, 16-fold for B/Victoria and 17-fold for A/Sichuan.

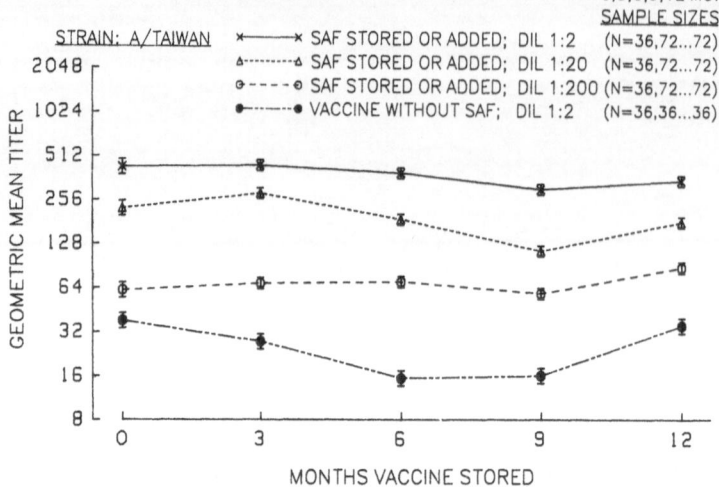

Fig. 7. Influenza vaccine stability study mouse potency tests.
Stability study. Trivalent influenza virus vaccine
(undiluted) was stored at 4°C in Tubex or vials with or
without SAF-m (1X) containing 2 mg/ml t-MDP. At time
intervals samples were assayed for Thimerosal concentration, pH and mouse potency. For the mouse potency
test, vaccine was diluted 1:2 in either vehicle or
concentrated adjuvant so that the amount of t-MDP and
vehicle was equivalent for both "stored" and "added"
samples. Further dilutions of 1:20 and 1:200 were made
in 1X SAF-m. Groups of eighteen seven-week-old female
mice were injected I.M. with 0.1 ml of preparation. Each
mouse received 1.75 μg HA of each strain at the 1:2
dilution and 100 ug t-MDP. Thirty-five days later serum
samples were assayed for HI antibody. Since the results
with Tubex and vials were equivalent statistically these
values were pooled for analysis.
Note. Standard error bars are based upon the pooled mean
square error.

Because of the large numbers of mice involved, these are our best estimates
of the true effect of undiluted adjuvant on this particular formula of
Wyeth-Ayerst influenza virus vaccine.

Safety

Fortunately, the components of SAF-m should be compatible with human
use. Tween 80 has been used in the Wyeth-Ayerst influenza vaccine and many
other products for a long time. The parent compound, MDP and derivatives of
it, have been used in several human studies (Telzak et al, 1986). Squalane
is a natural breakdown product of squalene on human skin and it has been
used as an adjuvant for animal vaccines and experimental human cancer vaccines (Vosika, 1983). Pluronic L121 has not been used in humans but other
Pluronics have been. Pluronic F-68 was used in humans as an emulsifier for
liver imaging (Clarke et al, 1985), as a component in artificcial blood
substitute (Vercellotti et al, 1982) and for other medical applications.

Wyeth-Ayerst is planning to conduct a requisite number of safety
studies to support clinical trials and registration. Three studies are
reported here.

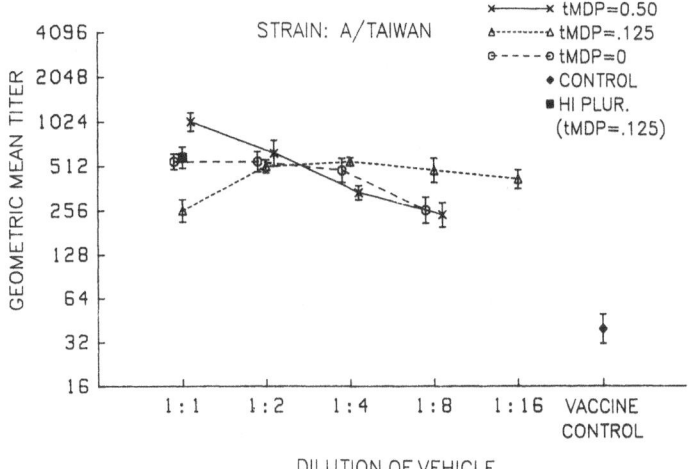

Fig. 8. Dilution of SAF-m vehicle in constant t-MDP. Groups of ten
CD-1 seven-week-old female mice were injected I.M. with 0.25
ml of influenza virus vaccine containing dilutions of vehicle
and various amounts of t-MDP. Each mouse received 3.75 μg
HA/strain. Twenty-eight days later, serum samples were assayed
for HI antibody.

Dilution Study. From a safety point of view it is prudent to use no
more of a compound in humans than is needed for immunization. It is also
our goal to be certain that there is no increase in local reactogenicity
(pain at the injection site) over the current baseline levels of standard
influenza virus vaccine. Therefore, we carried out a study in mice to
determine the effect of dilution of the adjuvant on efficacy. The vehicle
was diluted up to 16-fold while using three different concentrations of t-
MDP (Fig. 8). These HI results with the A/Taiwan component (H_3N_2) were
representative of those seen with the other components of the 1989-1990
vaccine (B/Yamagata and A/Shanghai). Diluting the vehicle 1:8 resulted in a
4-fold drop in titer using 0.5 mg t-MDP. The remaining titer was still 7-
fold better than non-adjuvanted vaccine. Using 0.125 mg or no t-MPDP
resulted in no significant change with dilution of vehicle as judged by the
Student-Newman-Keuls test at the 5% level. The mid-level concentration of
t-MDP showed a tendency to be superior to the other two t-MDP concentrations. These experiments were carried out using one-half the concentration
of the Pluronic L121 (1.25% in the undiluted vehicle) because other experiments had shown that the higher amount was unnecessary. One group of mice
in this experiment confirmed that fact. It appears, therefore, that there
could be considerable latitude to adjust the amount of vehicle employed in
humans based on the relatively flat vehicle dilution curves seen in this
experiment.

Guinea Pig Toxicity Study. A publication by Nagao et al (1985) gave us
some concern about the remote possibility that tuberculin positive individuals might react when injected with MDP analogs. Nagao injected M. tuberculosis (M.tb) in FIA into the footpads of guinea pigs. Three to six weeks
later he injected MDP intracutaneously into the "flank" and observed
swelling and necrosis in the footpads within 24 hours. This finding was
surprising to us in light of the fact that MDP is not itself immunogenic.

Table 4 shows our investigation of Nagao's model. We were able to
repeat Nagao's finding of increased necrosis and swelling in the left
footpads after a provocative injection in the left dorsal gluteus (groups 1
and 2). Necrosis occurred with two sources of MDP and with t-MDP combined

Table 4. Guinea Pig Footpad Experiment

Group	(#)	Description		Average Swelling* 2nd Injection		Average Necrosis 2nd Injection	
		1st Injection	2nd Injection	Before	24 Hours After	Before	24 Hours After
1	(6)	50µg MTb in FIA	1000µg MDP (Sigma)	2.8	3.7	0.8	3.5
2	(6)	50µg MTb in FIA	1000µg MDP (Syntex)	3.0	3.5	2.2	3.5
3	(6)	50µg MTb in FIA	100µg MDP (Sigma)	3.0	3.5	1.3	3.7
4	(6)	50µg MTb in FIA	1000µg tMDP	2.8	3.8	2.0	3.7
5	(6)	50µg MTb in FIA	100µg tMDP	3.2	3.5	1.7	2.2
6	(6)	50µg MTb in FIA	100µg butyl–abu–MDP	2.3	1.8	1.3	1.5
7	(6)	50µg MTb no FIA	1000µg tMDP	0.0	0.0	0.0	0.0
8	(6)	FIA only	1000µg tMDP	1.0	1.3	0.0	0.2
9	(8)	50µg MTP in FIA	PBS	3.3	3.3	2.6	2.8
10	(3)	50µg MTP in FIA (L hind leg)	1000µg tMDP	–	–	0.5[a] 0.0[b]	3.3[a] 0.0[b]
11	(3)	50µg MTP in FIA	1000µg tMDP + Vehicle + flu	3.0	4.0	2.7	4.0

*using a subjective grading on a scale of 0 – 4
[a]measurement of left hind leg
[b]measurement of footpad

with SAF-m and influenza vaccine (group 11). Comparison of groups 1, 2, 3, 4 and 5 using a subjective grading scale of 0 to 4 demonstrated that t-MDP at lower doses provoked less necrosis than MDP. This difference in necrosis was statistically significant at the 5% confidence level. Another Syntex compound, butyl-abu-MDP also had a reduced effect compared to MDP. As illustrated by group 10, necrosis occurred at sites other than the footpad. Most importantly, the reactions did not occur without FIA (groups 7 and 8). In fact, there was a trace of a reaction with FIA alone. This dependence on FIA strongly suggests that the phenomenon was artifactual in nature and may not present any risk to humans. Indeed, we noticed that the swelling and necrosis increased gradually in the unchallenged controls even without the provocative injection which seemed only to accelerate the reaction. Perhaps the MDP simply irritated the site to enhance the reaction caused by the FIA and M.tb. It seems unlikely that the phenomenon was immunological in nature especially since Nagao showed that tuberculin antigens in the form of purified protein derivative (PPD) did not elicit the reactions.

Nagao could not demonstrate the phenomenon in mice, rats or rabbits and the reaction was not reported in the limited clinical trials conducted by others with MDP and its analogs. These facts plus our investigation suggest that we should proceed to trials on humans. Due caution will certainly be exercised.

Toxicology. Some of the toxicology studies being carried out at Wyeth-Ayerst are described in Table 5. These studies were conducted under the exacting procedures of Good Laboratory Practices (GLP) outlined by the FDA. Animals were injected with adjuvant alone or with mixtures of Wyeth-Ayerst influenza vaccine and SAF-m. Histopathological examination revealed only minimal to moderate muscle irritation depending on the species and dilution of adjuvant. Dogs were more tolerant than rabbits with respect to intramuscular toxicities. Minimal muscle irritation was caused by the injection procedure alone. No persistent nodules were seen. Although the microscopic muscle examination is not complete, the gross observations appeared normal.

Table 5. Intramuscular Toxicology Studies

Animals: 1) New Zealand White Rabbits (60)
            2) Beagle Dogs (60)

All animals were of one sex (female).

Treatment Groups: 6 animals of each species in each group

1. Physiological saline
2. Influenza vaccine alone (marketed product)
3. 0 t-MDP, undiluted vehicle*, without influenza vaccine
4. 0 t-MDP, undiluted vehicle*, with influenza vaccine
5. 1.0 mg/dose t-MDP, undiluted vehicle, without influenza vaccine
6. 1.0 mg/dose t-MDP, undiluted vehicle, with influenza vaccine
7. 0.125 mg/dose t-MDP, undiluted vehicle, with influenza vaccine
8. 1.0 mg/dose t-MDP, 1:8 vehicle, with influenza vaccine
9. 0.125 mg t-MDP, 1:8 vehicle, with influenza vaccine
10. 1 mg t-MDP/dose alone in physiological saline

Duration: 14 days with sacrifice on day 15 for 3 of 6 animals/group. Remaining animals were sacrificed after a 4-week nondosing recovery period.

Treatment Regimen: One 0.5 ml intramuscular injection daily into the muscles of the upper hind limbs, alternating the sides (right and left) each day.

The rabbits and dogs had routine physical examinations: ophthalmologic examinations; hematology and clinical chemistry studies; and urinalyses. On day 15, 3 of 6 animals per group were subject to a complete necropsy that included all routine tissues, injection sites, and unusual lesions/masses. A similar necropsy was done on all survivors following a 4-week postdosing recovery period.

*low pluronic (1.25%) vehicle used in all adjuvants

CONCLUSION

Current influenza vaccines are reasonably effective but grossly under-utilized. The scientific community has expended a great effort to increase the efficacy of influenza vaccines. It is hoped that these efforts will come to fruition soon. New, more efficacious vaccines and widespread publicity should increase vaccine acceptance and reduce the tragic number of deaths caused by influenza virus.

Acknowledgements

We are indebted to Mary Hasson for assistance in the preparation of the manuscript and to Noelene Byars for helpful discussions.

REFERENCES

Allison, A.C. and Byars, N.E., 1986, An adjuvant formulation that selectively elicits the formation of antibodies of protective isotypes and of cell-mediated immunity, J.Immunol.Meth., 95:157.

Anon, 1989, Prevention and control of influenza: Part 1, Vaccines, JAMA, 261:3220.

Bachmayer, H., 1976, Split and subunit vaccines, in: "Influenza: Virus, Vaccines, Strategy", P. Selby, ed., Academic Press, N.Y.

Byars, N.E., Allison, A.C., Harmon, M.W. and Kendal, A.P., 1990, Enhancement
of antibody responses to influenza B virus hemagglutinin by use of a
new adjuvant formulation, Vaccine, 8:49.

Cate, T.R., Kasel, J.A., Couch, R.B., Six, H.R. and Knight, V., 1977,
Clinical trials of bivalent influenza A/New Jersey/76-A/Victoria/75
vaccines in the elderly, J.Infect.Dis., 136:S518.

Cate, T.R., Couch, R.B., Parker, P. and Baxter, B., 1983, Reactogenicity,
immunogenicity, and antibody persistence in adults given inactivated
influenza virus vaccines, Rev.Inf.Dis., 5:737.

Check,, W.A., 1984, Renewed control efforts emphasize "unrecognized"
influenza threat, JAMA, 251:2629.

Chen, K. and Quinnan, G.V., Jr., 1989, Secretory immunoglobulin A antibody
response is conserved in aged mice following oral immunization with
influenza virus vaccine, J.Gen.Virol., 70:3291.

Clarke, M.B., Tyrrell, D.A. and Barrett, J.J., 1985, Normal volunteer
studies with modified sup 9sup 9Tc(m) tin colloid, Nucl.Med.Commun.,
6:641.

Douglas, R.G., Jr., 1990, Prophylaxis and treatment of influenza,
N.Engl.J.Med., 322:443.

Fulk, R.V., Fedson, D.S., Huber, M.A., Fitzpatrick, R., Howar, B.F. and
Kasel, J.A., 1969, Antibody responses in children and elderly persons
following local or parenteral administration of an inactivated
influenza virus vaccine, A2/Hong Kong/68 variant, J.Immunol.,
102:1102.

Gross, P.A., Quinnan, G.V., Gaerlan, P.F., Denning, C.R., Davis, A.,
Lazicki, M. and Bernius, M., 1982, Potential for single high-dose
influenza immunization in unprimed children, Pediatrics, 70:982.

Gross, P.A., Quinnan, G.V., Jr., Weksler, M.E., Setia, U. and Douglas, R.G.,
Jr., 1989, Relation of chronic disease and immune response to
influenza vaccine in the elderly, Vaccine, 7:303.

Guink, N.E., Kris, R.M., Goodman-Snitkoff, G., Small, P.A., Jr. and Mannino,
R.J., 1989, Intranasal immunization with proteoliposomes protects
against influenza, Vaccine, 7:147.

Hilleman, M.R., Woodhour, A., Friedman, A., Weibel, R.E. and Stokes, J.,
Jr., 1972, The clinical application of adjuvant 65, Ann.Allergy,
30:152.

Hirabayashi, Y., Kurata, H., Funato, H., Nagamine, T., Aizawa, C., Tamura,
S., Shimada, K. and Kurata, T., 1990, Comparison of intranasal
inoculation of influenza HA vaccine combined with cholera toxin B
subunit with oral or parenteral vaccination, Vaccine, 8:217.

Hobson, D., 1973, The potential role of immunological adjuvants in influenza
vaccines, Postgrad.Med.J., 49:180.

Jennings, R., Potter, C.W., Massey, P.M.O., Duerden, B.I., Martin, J. and
Bevan, A.M., 1981, Responses of volunteers to inactivated influenza
virus vaccines, J.Hyg., 86:1.c

Kaplan, J.E., Katona, P., Hurwitz, E.S. and Schonberger, L.B., 1982,
Guillain-Barre' syndrome in the United States, 1979-1980 and 1980-
1981. Lack of an association with influenza vaccination, JAMA,
248:698.

Katz, S.L., 1985, "New Vaccine Development Establishing Priorities, Vol. I.
Diseases of Importance in the United States", National Academy Press,
Washington, D.C.

Kurland, L.T., Wiederholt, W.C., Kirkpatrick, J.W., Potter, H.G. and
Armstrong, P., 1985, Swine influenza vaccine and Guillain-Barre'
syndrome, epidemic or artifact? Arch.Neurol., 42:1089.

Levine, M., Beattie, B.L. and McLean, D.M., 1987, Comparison of one- and
two-dose regimens of influenza vaccine for elderly men,
Canad.Med.Assoc.J., 137:722.

Lovgren, K., 1988, The serum antibody response distributed in subclasses and
isotypes after intranasal and subcutaneous immunization with influenza virus immunostimulating complexes, Scand.J.Immunol., 27:241.

Maassab, H.F., Heilman, C.A. and Herlocher, M.L., 1990, Cold-adapted influenza viruses for use as live vaccines for man, in: "Viral Vaccines, Advances in Biotechnological Processes", A.M. Izrahi, ed., Wiley, Liss, N.Y.

Margolis, K.L., Nichol, K.L., Poland, G.A. and Pluhar, R.E., 1990, Frequency of adverse reactions to influenza vaccine in the elderly, a randomized, placebo-controlled trial, JAMA, 264:1139.

Masihi, K.N., Lange, W., Brehmer, W. and Ribi, E., 1986, Immunobiological activities of nontoxic lipid A: Enhancement of nonspecific resistance in combination with trehalose dimycolate against viral infection and adjuvant effects, Int.J.Immunopharmac., 8:339.

Matzkin, H. and Nili, E., 1984, Accidental tenfold overdose of influenza vaccine: A clinical and serological study, Israel J.Med.Sci., 20:411.

McLaren, C., Williams, M.S., Bozeman, F.M., Mayner, R.E., Grubbs, G.E., Bartholow, W.E., Staton, E. and Ennis, F.A., 1978, Comparative antigenicity and immunogenicity of 1976 influenza virus vaccines: Results of mouse protection experiments, J.Biol.Stand., 6:315.

Miles, R.N., Potter, C.W., Clark, A. and Jennings, R., 1981, Reactogenicity and immunogenicity of three inactivated influenza virus vaccines in children, J.Biol.Stand., 9:379.

Mostow, S.R., 1986, Influenza - a preventable disease not being prevented, Am.Rev.Respir.Dis., 134:1.

Mostow, S.R., Schoenbaum, S.C., Dowdle, W.R., Coleman, M.T. and Kaye, H.S., 1973, Inactivated vaccines. I. Volunteer studies with very high doses of influenza vaccine purified by zonal ultracentrifugation, Postgrad.Med.J., 49:152. T

Murray, R., Cohen, P. and Hardegree, M.C., 1972, Mineral oil adjuvants: Biological and chemical studies, Ann.Allergy, 30:146.

Nagao, S. and Tanaka, A., 1985, Necrotic inflammatory reaction induced by muramyl dipeptide in guinea pigs sensitized by tubercle bacilli, J.Exp.Med., 162:401.

Parkman, P.D., Galasso, G.J., Top, F.H. and Noble, G.R., 1976, Summary of clinical trials of influenza vaccine, J.Infect.Dis., 134:100.

Pietrobon, P.J., Popescu, M.C., Hyde, A.M. and Recine, M.S., 1989, Development of a novel liposome formulation for use as vaccine adjuvant, in: "Modern Approaches to New Vaccines" (Abstracts), F. Brown, R. Chanock, H.S. Ginsberg and R. Lerner, eds., Cold Spring Harbor, N.Y.

Powers, R.D., Hayden, F.G., Samuelson, J. and Gwaltney, J.M., Jr., 1984, Immune response of adults to sequential influenza vaccination, J.Med.Virol., 14:169.

Prevost, J.M., Petermans, J., Lamy, F. and Huygelen, C., 1973, Immune response to vaccination with a live influenza virus (H3N2) vaccine ("Ann" strain), Inf.and Immun., 8:420.

Quinnan, G.V., Schooley, R., Dolin, R., Ennis, F.A., Gross, P. and Gwaltney, J.M., 1983, Serologic responses and systemic reactions in adults after vaccination with monovalent A/USSR/77 and trivalent A/USSR/77, A/Texas/77, B/Hong Kong/72 influenza vaccines, Rev.Inf.Dis., 5:748.

Salk, J. and Salk, D., 1977, Control of influenza and poliomyelitis with killed virus vaccines, Science, 195:834.

Salk, J.E., Pearson, H.E., Brown, P.N. and Francis, T.F., Jr., 1944, Protective effect of vaccination against induced influenza B, Proc.Soc.Exper.Biol.Med., 55:106.

Schoenbaum, S.C., 1987, Economic impact of influenza, the individual's perspective, Am.J.Med., 82:26.

Sears, S.D., Clements, M.L., Betts, R.F., Maassab, H.F., Murphy, B.R. and Snyder, M.H., 1988, Comparison of live, attenuated H1N1 and H3N2 cold-adapted and avian human influenza A reassortant viruses and inactivated virus vaccine in adults, J.Inf.Dis., 158:1209.

Stuart-Harris, C.H., 1969, Adjuvant influenza vaccines, Bull.WHO, 41:617.

Stuart-Harris, C., 1980, The present status of live influenza virus vaccine, J.Infect.Dis., 142:784.

Sullivan, K.M., Monto, A.S. and Foster, D.A., 1990, Antibody response to inactivated influenza vaccines of various antigenic concentrations, J.Inf.Dis., 161:333.

Telzak, E., Wolff, S.M., Dinarello, C.A., Conlon, T., Kholy, A.E., Bahr, G.M., Choay, J.P., Morin, A. and Chedid, L., 1986, Clinical evaluation of the immunoadjuvant murabutide, a derivative of MDP, administered with a tetanus toxoid vaccine, J.Inf.Dis., 153:628.

Thibodeau, L., Naud, P. and Boudreault, A., 1981, An influenza immunosome: Its structure and antigenic properties. A model for a new type of vaccine, in: "Genetic Variation Among Influenza Viruses", D.P. Nayak, ed., Academic Press, Inc., New York.

Vercellotti, G.M., Hammerschmidt, D.E., Craddock, P.R. and Jacob, H.S., 1982, Activation of plasma complement by perfluocarbon artificial blood: Probable mechanism of adverse pulmonary reactions in treated patients and rationale for corticosteroid prophylaxis, Blood, 59:1299.

Vosika, G.J., 1983, Clinical immunotherapy trials of bacterial components derived from Mycobacteria and Nocardia, J.Biol.Resp.Modif., 2:321.

Waldman, R.H. and Coggins, W.J., 1972, Influenza immunization: Field trial on a university campus, J.Inf.Dis., 126:242.

Waldman, R.H., Bergmann, K., Stone, J., Howard, S., Chiado, V., Jacknowitz, A., Walkman, E.R. and Khakoo, R., 1987, Age-dependent antibody response in mice and humans following oral influenza immunization, J.Clin.Immunol., 7:327.

Wright, P.F. and Karzan, D.T., 1987, Live attenuated influenza vaccines, Prog.Med.Virol., 34:70.

NONIONIC BLOCK POLYMER SURFACTANTS AS ADJUVANTS IN VACCINES

Guy J.W.J. Zigterman*, Andre F.M. Verheul** and Harm Snippe**

*Department of Bacteriology Research
Intervet International, Boxmeer
**Eykman-Winkler Laboratory of Medical Microbiology
Utrecht University
Utrecht, The Netherlands

SCOPE OF THIS REVIEW

Nonionic Block Polymers (NBPs) are simple copolymers of polyoxyethylene (POE) and hydrophobic polyoxypropylene (POP) and differ in molecular weight, percentage POE and the mode of linkage of POE abd POP-groups. These adjuvants are currently investigated on their potential use in vaccines. The mode of action of these adjuvants is largely unknown (Hunter et al, 1981, 1989; Byars and Allison, 1987). It is difficult to hypothesize that differences in biologic activities among NBPs could be due to the presence of active sites on one molecule that are not present on all of them. The experiments described in this review point out that NBPs are adjuvants which interfere with components of the immune system while interactions with the antigen are also noticed. The models put forward by Hunter and Allison on the mechanism by which pluronic polyols interfere with the immune system are presented. Finally, our efforts to develop a semi-synthetic pneumococcal vaccine are discussed and the effects of NBPs on the antibody avidity and isotype distribution are described.

SURFACTANTS AND NONIONIC BLOCK POLYMERS

During the last decades the development of subunit vaccines was hampered to a large extent by the reduced immunogenicity of these subunit vaccines compared to the original whole virus or bacterium vaccines which were used either as (attenuated) living organisms or inactivated organisms. This reduced immunogenicity of subunit vaccines necessitated the use of adjuvants to reach a desired status of immunity in vaccinated animals.

Among the candidates to be incorporated as adjuvant in vaccines, the surfactants form a special group (reviewed recently by Hilgers et al, 1989). Surfactants, due to the presence of hydrophobic and hydrophilic regions (Cahn and Lynn, 1983; Griffin, 1979) are able to have interactions with both hydrophobic molecules on the one hand and hydrophilic molecules on the other. For this reason they are able to stabilize oil-in-water and water-in-oil emulsions. These, as well as other surface active properties are related to the solubility in water as expressed in the HLB value (Table 1). The HLB value can be assessed by using one of the following formulas:

Table 1. HLB Value of Surfactants in Relation to Their Solubility in Water and Applications (Cahn and Lynn, 1983; Griffin, 1979)

HLB range	Solubility in Water	Application
< 3	No dispersibility	Spreading agent
3-6	Poor dispersibility	Water-in-oil emulsion
6-8	Milky dispersion	Wetting agent
8-10	Stable milky dispersion	Oil-in-water emulsion
10-13	Translucent to clear dispersion	Oil-in-water emulsion
>13	Clear solution	Solubilization/detergent

Table 2. Chemical Properties of NBP Surfactants

Compound	Average[a] molecular weight	% POE[a]	Hydrophile-lipophile balance[b]	Structure
Triblock				
L72	2750	20	–	
L81	2750	10	2.0	
L92	3650	20	5.5	POE-POP-POE
L101	3800	10	1.0	
L121	4400	10	0.5	
L122	5000	20	4.0	
Reversed Triblock				
25R1	2800	10	2.3	POP-POE-POP
31R1	3200	10	1.7	
Octablock				
T1101	5600	10	2.0	
T1301	6800	10	1.5	POE-POP $_X$ POP-POE
T1501	7900	10	1.0	POE-POP ^{X} POP-POE
Reversed Octablock				
130R1	6800	10	1.4	POP-POE $_X$ POP-POP
130R2	7740	20	2.9	POP-POE ^{X} POE-POP

[a]Average molecular weight and % POE according to the manufacturer.
[b]Hydrophile-lipophile balance-values were obtained from Hunter and Bennett (1984).

HLB = 20 (1-S/A), where S is the saponification number of the ester and A is the acid number of separated acid, or HLB = E/5, where E is the weight percentage ethyleneoxide content.

Nonionic block polymer surfactants (NBPs) belong to the group of surfactants and have different physicochemical properties as a funciton of their chemical composition (Table 2). As is evident from the series of triblocks in Table 2 (L72 to L122), the influence of the hydrophilic polyoxyethylene (POE) groups on the HLB value is most pronounced when the total molecular weight (i.e. the sum of the POE and polyoxypropylene (POP) groups) is smallest. The normal triblocks differ from the reversed triblocks by the linkage of POE to POP groups and therefore by the accessibility of various parts of the molecule for interaction with other molecules. The octablocks and reversed octablocks are mutually related to each other like the triblock and the reversed triblock but both octablocks have a more complex chemical structure than the triblocks.

All three parameters (percentage POE, molecular weight and structure) have an influence on the adjuvant activity as will be discussed below.

DEMONSTRATION OF ADJUVANT ACTIVITY OF NBPs

At the beginning of the eighties it was demonstrated that NBPs were able to stimulate various immune responses. Modulation of anti-bovine serum albumin (BSA) antibody production was segregated from granuloma-formulation when oil-in-water emulsions containing BSA and various NBPs were used (Hunter et al, 1981).

The hypothesis that adjuvant activity was inversely correlated with HLB-value (i.e. strong adjuvant activity was correlated with low HLB-values) later proved to be an underestimation of the complexity of adjuvant activity. On the other hand, it was demonstrated that triblock L121 not only stimulated the humoral anti-BSA but also the humoral anti-sheep erythrocyte response (Snippe et al, 1981). Moreover, triblock L101 was shown to be a potent adjuvant for the induction of a cellular immune response as was manifested by the induction of delayed type hypersensitivity with this NBP (Snippe et al, 1981).

Adjuvant activity in the case of NBPs appeared to be complex: while with the same antigen one NBP stimulated antibody production another NBP could stimulate cellular immune responses. When different antigens were compared the situation was even more complex. In the case of BSA as antigen relatively large NBPs with POE tails flanking POP centres and relatively low amounts of POE, were the better adjuvants (1101, L101, L121, T1505 in increasing order) (Hunter and Bennett, 1984). The idea that HLB-values determined adjuvant activity was abandoned: the ability to form adsorptive surfaces appeared to be involved in adjuvant activity. This however is not a general rule. Although our later experiments with oligosaccharide-protein conjugates were not in conflict with this hypothesis (see below), the experiments with dinitrophenyl-alanyl-glycyl-glycine (hapten J) as a hapten necessitated an adaptation of this hypothesis.

NBPs AND HAPTENATED ANTIGENS

When the immunomodulating activity of NBPs was tested in mice using liposomes haptenated with hapten J conjugated to phosphatidylethanolamine as antigen, the immunomodulating activity could be demonstrated with all NBPs tested (Table 3).

The most striking result was that NBPs 25R1 and 31R1 were very potent adjuvants for the immune response against haptenated liposomes while these compounds only marginally stimulated the immune response against BSA (Hunter and Bennett, 1984). Further experiments indicated that the height of the induced humoral immune response was correlated with the dose of NBP given (Fig. 1).

NBPs 31R1 and 25R1 appeared to be belonging to the stronger adjuvants for liposomes haptenated with J-PE: not only the immune response against a small dose of J-PE in liposomes, but also the immune response against an

Table 3. Influence of various NBPs on PFC-number and HA-titer

Block polymer	PFC/spleen x 10^{-3} [a,b]	HA-titer[a,b]
-	1.4 0.5 (5)	2.3 0.7 (10)
1101	10.4 1.0 (5)	4.3 1.0 (10)
1301	29.9 2.9 (5)	7.4 0.7 (10)
1501	21.3 1.9 (4)	7.0 1.0 (10)
L72	38.0 6.6 (5)	7.5 0.4 (10)
L81	36.6 4.3 (4)	8.1 0.5 (10)
L92	36.3 4.0 (5)	7.8 0.5 (10)
L101	17.1 2.4 (4)	5.7 1.1 (10)
L121	13.8 2.6 (5)	6.0 0.9 (10)
L122	12.4 1.0 (4)	6.0 1.0 (10)
25R1	33.7 5.7 (5)	8.0 0.5 (10)
13R1	46.2 8.4 (5)	8.2 0.4 (10)

[a]Mice were immunized i.p. on day 0 with J-PE liposomes (containing 0.2 nmol J-PE) and 200 nmol of NBP. On day 4 the number of splenic PFC and the HA titer in serum were determined. Results are expressed as means standard deviations with the number of mice used in brackets.

[b]Significance of difference between test groups and control groups receiving J-PE liposomes only, was determined with Student's t-test. All groups receiving J-PE liposomes and NBPs had significantly higher ($p < 0.001$) numbers of PFC and HA titers than the control group.

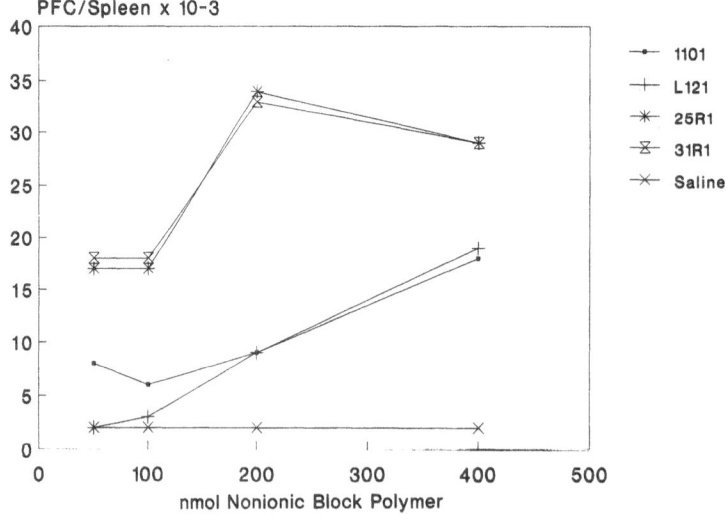

Fig. 1. Influence of adjuvant dose on adjuvant activity. Mice (n=10 in each group) received i.p. simultaneously 0.2 nmol J-PE incorporated into liposomes and varying amounts of NBPs (50-400 nmol) on day zero. PFC numbers in the spleen were determined on day 4.

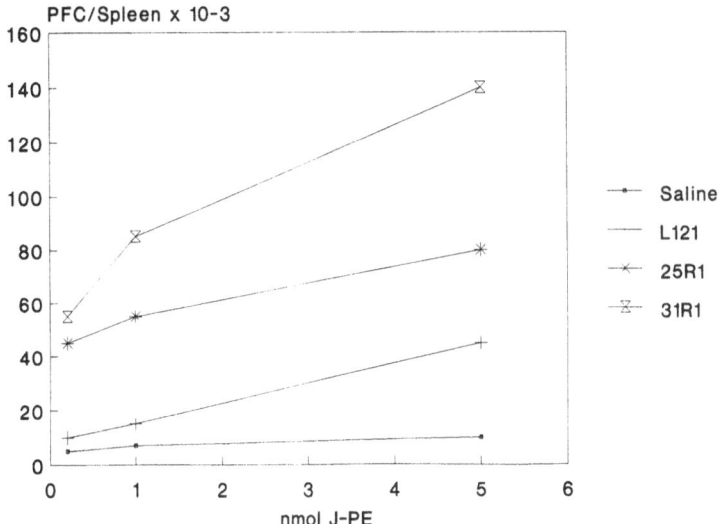

Fig. 2. Influence of antigen dose on adjuvant activity. Mice
(n=10-15 in each group) received i.p. simultaneously
200 nmol NBP and varying amounts of J-PE (0.2-5.0 nmol)
incorporated into liposomes on day zero. PFC numbers in
the spleen were determined on day 4. Mice were immunized
with the indicated dose of J-PE incorporated into liposomes with or without NBP polymer addition.

optimal dose of J-PE (i.e. 5 nmol) was stimulated efficiently by simultan-
eous administration of NBP 25R1 or 31R1 (Fig. 2). For the detection of
adjuvant activity the NBPs, however, did not necessarily have to be
administered simultaneously with the antigen. In contrast, when 25R1 was
administered at various intervals before or after immunization with J-PE
liposomes (Fig. 3), activity was detected at all intervals. The optimal
time of 25R1 administration, however, proved to be six h before immunization. This, together with the fact that lower doses of NBP administered
simultaneously with J-PE liposomes resulted in lower immune responses (Fig.
1) led to the hypothesis that local interactions were responsible for the
increased adjuvant activity observed after pretreatment with NBP 25R1.

Experiments in nude mice indicated that T-cells were not of major
importance for adjuvant activity of NBPs since similar adjuvant activity was
detected in controls and nude mice (data not shown). NBPs do not belong to
the group of B-cell mitogens with adjuvant activity. NBPs L121, T1101, 31R1
and 130R2 did not stimulate proliferation of B-cells (Zigterman, 1988).
Moreover, LPS-induced proliferation of splenic lymphocytes was inhibited by
these NBPs. The fact that pretreatment with NBP 25R1 6 h before immunization enhanced the humoral immune response indicated that the adjuvant
activity probably was exerted via interactions with elements involved in the
early induction stage of the immune response. Most prominent candidates for
interaction with NBPs are the complement system and phagocytic cells which
are both involved in the induction stage of the immune response and are both
susceptible to immunomodulators (Behbehani et al, 1985; Biozzi et al, 1985;
Egwang and Befus, 1984; Kido et al, 1984; Klerx et al, 1986; Loos and
Bitter-Suermann, 1976; Pepys, 1974; Unanue and Allen, 1987). Although
Hunter and Bennett (1984) reported activation of the human complement system
by NBPs, we were not able to repeat their results with mouse serum as
complement source (Fig. 4). When liposomes labelled with carboxyfluorescein
(CF) were incubated in either normal or heat-inactivated mouse serum, the
spontaneous release of this internal label as observed in buffer was

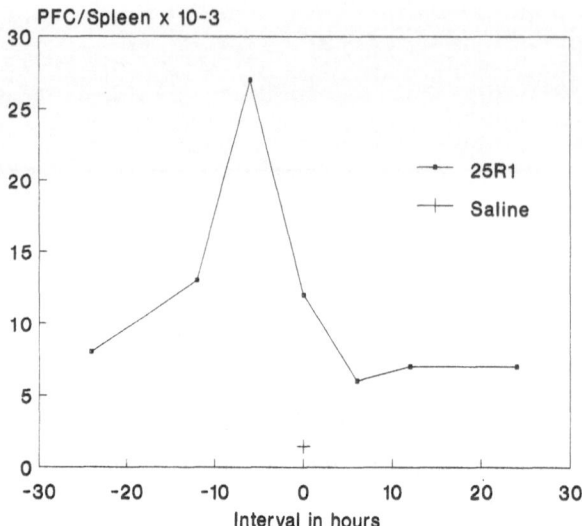

Fig. 3.   Influence of the interval between administration of 25R1
          and antigen on the immune response. Groups of mice (n=10)
          were injected i.p. with 200 nmol 25R1 and J-PE liposomes
          (containing 0.2 nmol J-PE) either simultaneously or separ-
          ated by different intervals. PFC numbers were determined
          on day 4.

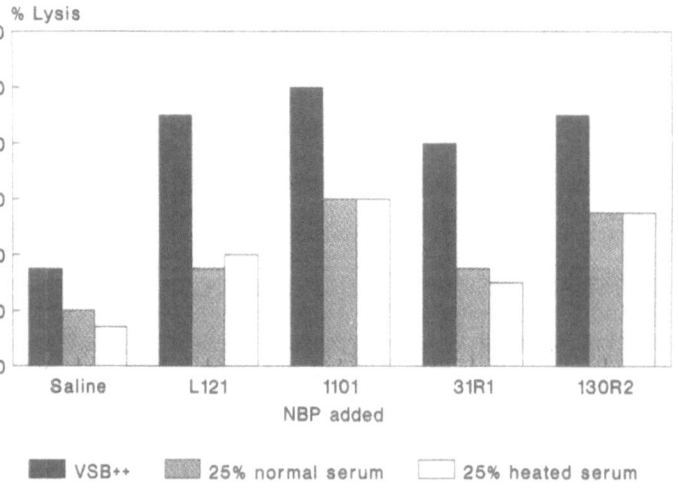

Fig. 4.   Influence of NBPs on lysis of liposomes in buffer and
          serum. After preincubation during 30 min at 39°C of NBP
          in VSB^{++}, or 25% normal mouse serum or 25% heat inactiv-
          ated mouse serum, liposomes were added and incubated for
          another hour at 39°C. Carboxyfluorescein-release was
          assayed in triplicate in two experiments.

reduced. Moreover, the increased release of label due to the action of NBPs was also counteracted by addition of serum. This stabilizing effect of serum addition was similar in normal and heat-inactivated serum. Therefore a role of complement in the adjuvant activity of NBPs was unlikely. This was confirmed later in mice treated with cobra venom factor in which adjuvant activity of NBPs was indistinguishable from that observed in control mice (Fig. 5).

The activity of peritoneal cells was modulated by pretreatment with NBPs. When phagocytosis of liposomes was measured as parameter of phagocyte activity, the NBPs tested all inhibited the phagocytic activity of peritoneal cells. To investigate the precise interactions between NBPs and phagocytic cells more experiments have to be performed. That phagocytic cells may play a crucial role in NBP adjuvant activity was also indicated by the recent results of Hunter et al (1989) and Howerton et al (1990) who demonstrated that NBPs are able to enhance Ia expression on peritoneal phagocytic cells.

In conclusion it can be stated that various mechanisms may be involved in immunomodulating activity of NBPs. Although several mechanisms are suggested, until now all these are concerned with the induction stage of the immune response. As was evident from studies with oil-in-water emulsions containing protein antigens, NBPs were thought to be involved in localization of antigen on the surface of the oil-spheres (Allison and Byars, 1986; Hunter and Bennett, 1986, 1987), where in some cases modulation of complement activation may occur.

Due to their amphiphilic character the NBPs may focus antigen on the membrane of antigen presenting cells (Hunter and Bennett, 1987). Processing of the antigen by these cells may then be influenced in such a way that immunogenic determinants are presented on the surface when Ia expression enables optimal presentation. Probably a complex interference with all

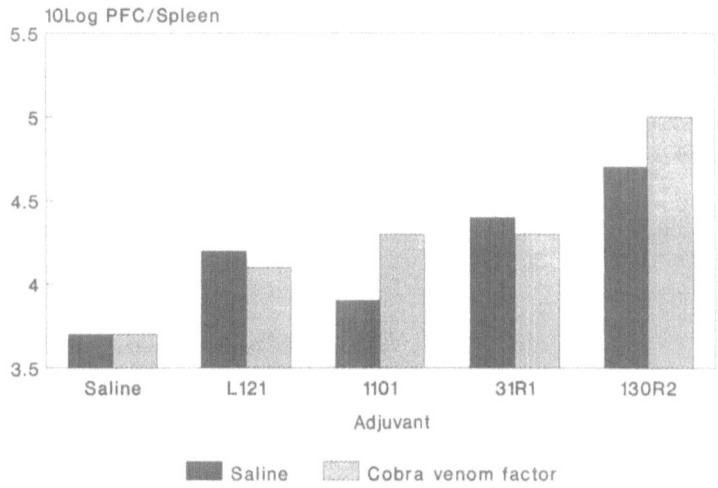

Fig. 5. Adjuvant activity of NBPs in mice treated with cobra venom factor. Groups of mice (n=10) were used as such or injected twice with 20 units of cobra venom factor prior to immunization with J-PE liposomes (containing 0.1 nmol J-PE) and 200 nmol of the indicated NBPs. ^{10}Log PFC numbers/spleen were determined four days after immunization.

these processes is responsible for the promotion of the immune response. When different antigens are used, different processes or parts of the processes described above seem to be important and therefore adjuvant requirements differ from antigen to antigen as demonstrated in Table 4. Of the antigens shown here, BSA and HS-BSA are more or less similar as the immune response is not stimulated by the reversed triblock 31R1. Both BSA and liposomes haptenated with hapten J, on the other hand, were combined efficiently with 31R1 resulting in strong immune responses. Although the causes of these different adjuvant requirements are not understood completely yet, it is evident that the immune response against several physicochemically different antigens is stimulated efficiently with one of the NBPs.

PNEUMOCOCCAL OLIGO- AND POLYSACCHARIDE PROTEIN CONJUGATES: MODULATION OF VARIOUS ASPECTS OF THE IMMUNE RESPONSE BY NBPs

The process of vaccine development, in particular the search for successful adjuvants, comprises not only the screening of antibody titers after vaccination but also other aspects of major importance. Although the induction of protection is often reached by increased titers, other factors as well may prove to be of major importance: distribution of antibodies over the various isotypes, extension of the response to hitherto silent B-cell clones, increase of avidity and changing of the ontogeny of the immune response. These beneficial aspects should not be accompanied by negative aspects such as pyrogenicity and local tissue reaction or other unwanted side effects.

The accepted candidates for vaccines against encapsulated pathogens are capsular polysaccharide-protein conjugates or capsular polysaccharide derived oligosaccharide-protein conjugates (Dick et al, 1989). In general, two types of conjugates are distinguished: (1) neoglycoproteins and (2) lattice conjugates. In a neoglycoprotein conjugate the protein carrier is substituted by a number of oligosaccharide chains having single attachment points, leading to good soluble conjugates. When each reaction partner however has multiple points for covalent attachment, conjugation will form a cross-linked grid or lattice matrix, linking the components together (lattice conjugate) (Dick et al, 1989). As an example of a neoglycoprotein conjugate, oligosaccharides derived from S. pneumoniae type 3 were coupled to the protein carriers BSA and KLH by the method of Svenson and Lindberg (1979). A capsular polysaccharide type 14-BSA conjugate was used as a model for a lattice conjugate (Verheul et al, 1989).

The IgG antibody levels induced by combinations of HS-KLH or HS-BSA with either NBP L121 or NBP T1501 were higher than the titers without NBP (Table 5). This increase in total IgG was a reflection not only of increased production of the predominant IgG1 isotype (Table 6) but also of increased production of the other isotypes. In the case of HS-BSA the IgG1 and IgG2a responses were stimulated to a large extent by NBP L121 while NBP 1501 stimulated IgG1, IgG2b and IgG3 responses. With HS-KLH as antigen, responses of all isotypes were stimulated by both NBPs. For the S14PS-BSA conjugate, similar results were obtained. Immunization with the conjugate alone resulted in the induction of IgG antibody levels but not all IgG subclasses were present (Tables 5 and 7). IgG1 was the dominant subclass while no IgG2a antibodies were induced. L121 not only enhanced the antibody levels of isotype already present, but was also able to induce IgG2a antibodies (Table 7). All this is in agreement with findings of Allison and Byars (1986) who reported the stimulation of IgG2a antibody responses in animals vaccinated with L121-containing vaccines. Furthermore, addition of CFA also increased the total IgG antibody levels and resulted in the induction of all IgG subclasses (Table 7).

Table 4. Adjuvant Activity of Various Non-ionic Block Polymers for the Humoral Response to Four Antigens[a]

Adjuvant type	HLB[b]	Structure[b]	stimulation of antibody response to			
			BSA[c]	J-BSA	HS-BSA	J-PE
L81 triblock	2.0	spherical drops	no	moderate	no	strong
L101 triblock	1.0	fibres	moderate	moderate	moderate	moderate
L121 triblock	0.5	fibres	moderate	moderate	moderate	moderate
31R1 reversed triblock	1.7	spherical drops	no	strong	no	strong
T1101 octablock	2.0	oily fibres	moderate	moderate	moderate	poor
T1501 octablock	1.0	fibres	strong	moderate	moderate	strong

[a] The NBP were immobilized on oil droplets and mixed with the antigens BSA, J-BSA and HS-BSA. J-PE was incorporated in liposomes and mixed with NBPs. Mice were injected intraperitoneally and antibody titres were measured in serum.

[b] HLB values and structure of the NBP were obtained from (Hunter and Bennett, 1986).

[c] BSA, bovine serum albumin; J-BSA, dinitrophenyl-alanyl-alanyl-glycine (J) coupled to BSA (J:BSA ratio is 22); J-PE, J-phosphatidylethanolamine incorporated in liposomes; HS-BSA, hexasaccharide fragment of the capsular polysaccharide of Streptococcus pneumoniae type 3, coupled to BSA (HS:BSA ratio is 7).

Table 5. Modulation of HS-BSA and HS-KLH Induced IgG Responses by NBPs

Adjuvant	Time	Antigen[a]		
		HS-BSA[c]	HS-KLH[c]	S14PS-BSA[d]
–	0	–3.6	–3.6	–4.0
none	14/2[b]	–1.8	–0.2	0.3
	14/3	–1.6	–0.7	ND
T1501	14/2	–1.1	0.7	ND
	14/3	+0.2	0.9	ND
L121	14/2	–1.1	0.2	1.2
	14/3	–0.5	0.1	ND

[a]The antibody level was determined in the pool of sera of five
female (CBA/N x BALB/c) F1 mice after i.p. immunization with
50 µg HS-BSA or 50 µg HS-KLH with or without 200 nmol of the
indicated NBP in phosphate buffered saline.
[b]14/2: 14 days after secondary vaccination.
14/3: 14 days after tertiary vaccination.
0: preimmune.
[c]Titers are expressed as effective dose (ED) values i.e. ^{10}log
(titer test pool) $^{-10}$log (titer reference serum) using an S3
specific reference serum.
[d]Results are expressed as ED values using an S14 specific reference serum.

With regard to the recruitment of silent B-cell clones by certain
adjuvants, not many data are available yet. This was monitored in (CBA/N x
BALB/c) F1 female mice after several vaccinations with S. pneumoniae type 3-
derived oligosaccharides linked to KLH either with or without NBPs. Sera of
mice were analyzed by isoelectric focusing five days after the second (day
26) and third (day 47) vaccination. Antibodies directed against the oligosaccharides were detected by staining with oligo-saccharide-bovine-γ-glob-
ulin conjugated to peroxidase. The isoelectric focusing patterns revealed
relatively weak staining of bands when sera of mice vaccinated without NBP
were analyzed (Fig. 6). The intensity of bands was increased by the use of
NBPs L101, L121, 1101 or 1501. The patterns of bands observed in the sera
of the various test groups at the different times, revealed a large contin-
uity among individuals. NBPs in general increased the density of already
existing bands (which are hardly visible in the photograph) and minor
induction of new clones (e.g. bands observed at pH values > 8 when NBPs 1101
or 1501 were used). This is similar to modulation of immune responses by
Freund's complete adjuvant which also realized an enhanced secretion of
antibodies by already activated B-cells leading to an accelerated expansion
of the band pattern observed by isoelectric focusing (Nakashima and
Kamikawa, 1984). Measurement of antibody avidity in antisera by a competition ELISA test revealed that NBPs L121 and 1501 were able to prevent
decreased affinities of IgG2a antibodies observed after repeated vaccination
without NBPs (Fig. 7) (Dam et al, 1989a, b).

Similar results were obtained when S14PS-protein conjugates were used
as antigen. L121 enhanced the relative avidities of the evoked IgG antibodies to a level above those obtained after immunization with conjugate
alone or in combination with CFA (Table 8).

Table 6. IgG Isotypes in Serum After Immunization with HS–Protein Conjugates and NBPs

Conjugate + adjuvant	Time	ED value of antibody isotype[a]			
		IgG1	IgG2a	IgG2b	IgG3
Preimmune	0[b]	-3.92	-3.32	-2.79	-3.09
HS–BSA	14/2	-1.19	-2.09	-1.42	-2.06
	14/3	-0.97	-2.23	-1.34	-2.19
HS–BSA +1501	14/2	-0.54	-0.90	-0.29	-1.05
	14/3	0.74	-0.95	0.21	-0.29
HS–BSA +L121	14/3	0.51	-1.11	-0.89	-1.50
	14/3	0.32	-0.57	-0.85	-1.86
Preimmune		-3.92	-3.32	-2.79	-3.09
HS–KLH	14/2	-0.29	-1.48	-0.96	-0.41
	14/3	0.54	-2.10	-1.64	-1.08
HS–KLH +1501	14/2	0.76	-0.34	0.17	0.15
	14/3	0.93	-0.03	0.43	0.33
HS–KLH +L121	14/2	0.22	-0.18	-0.06	0.12
	14/3	0.27	-0.53	-0.34	0.18

[a]Sera of five female mice immunized as described in Table 5.
[b]14/2:14 days after the second immunization (day 35)
14/3:14 days after the third immunization (day 56)
   0:preimmune.

Table 7. Influence of Adjuvants on the IgG Subclass Distribution after Immunization with S14PS–BSA Conjugate

	S14PS–BSA (µg/ml)		+L121 (µg/ml)		+CFA (µg/ml)	
IgG1	70	20	460	200	620	200
IgG2a	0		120	80	20	10
IgG2b	5		60	20	30	10
IgG3	5		180	20	120	90

Groups of 5-6 ICR mice (outbred mouse strain, Charles River Laboratories, Raleigh, NC, USA) were immunized with either 38.4 µg of conjugate containing 20.8 µg of protein and 17.6 µg of S14PS) or in combination with L121 or CFA. Data are shown of sera obtained two weeks after the first booster injection.

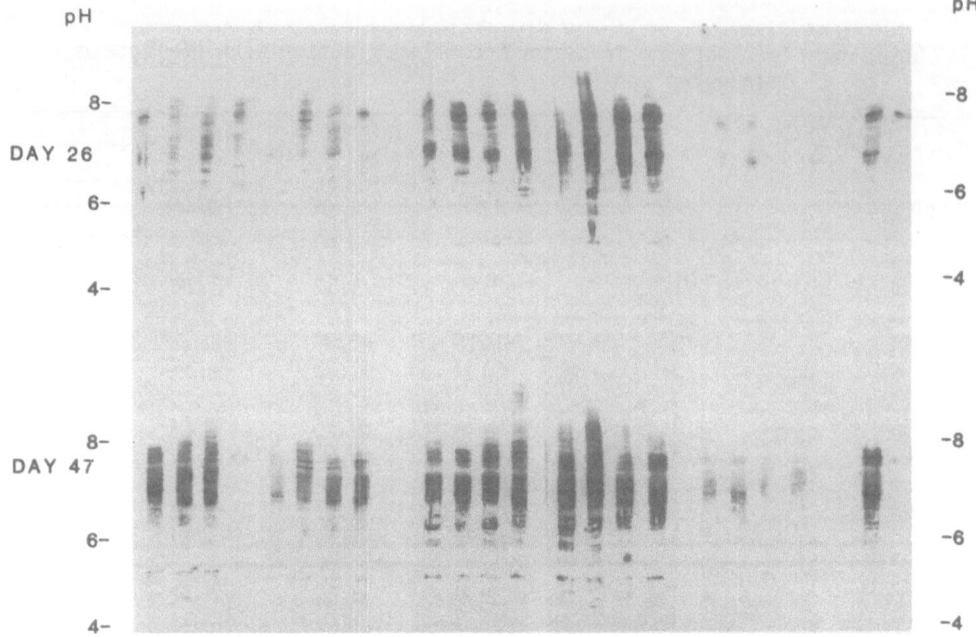

Fig. 6. Isoelectric focusing (IEF) patterns of female (CBA/N x
BALB/c) F_1 mice immunized i.p. with 50 μg HS-KLH with or
without 200 nmol NBP. From various test groups 3-4 mice
were analyzed by IEF 5 days after second and third immunization (day 26 and day 47). Mice were immunized with
HS-KLH and the NBPs L101, L121, 1101 and 1501 (-: mice
immunized with HS-KLH without NBP). H: hyperimmune refer
ence serum; P: Preimmune reference serum.

Modulation of the ontogeny of the immune response by NBPs was analysed
in young mice. When the BALB/c mice were immunized with HS-BSA and NBP 1501
or L101 at various ages, the immune response could be enhanced significantly
by both NBPs. Especially in mice immunized at the age of seven or 10 days,
the addition of L101 but more so 1501 resulted in increased IgM and IgG
titers and also increased protection against a lethal challenge with the
homologous pneumococcus type (Fig. 8). Considering the local reactions
observed after vaccination with NBP-containing vaccines not much research
has been done except for the experiments by Hunter on granuloma formation

Table 8. Relative Avidity Values of Sera of Mice Immunized with
S14PS-BSA Conjugate and Adjuvants

	IgG	IgG1	IgM
S14PS-BSA	2.2	2.3	2.6
S14PS-BSA + L121	3.2	3.5	3.1
S14PS-BSA + CFA	2.2	1.9	3.8

The relative avidity values are expressed as $^{-10}\log[I]_{50}$ values,
in which the $[I]_{50}$ is the concentration of capsular S14PS necessary
to decrease the amount of antibody bound to the Ag-coat by 50%.

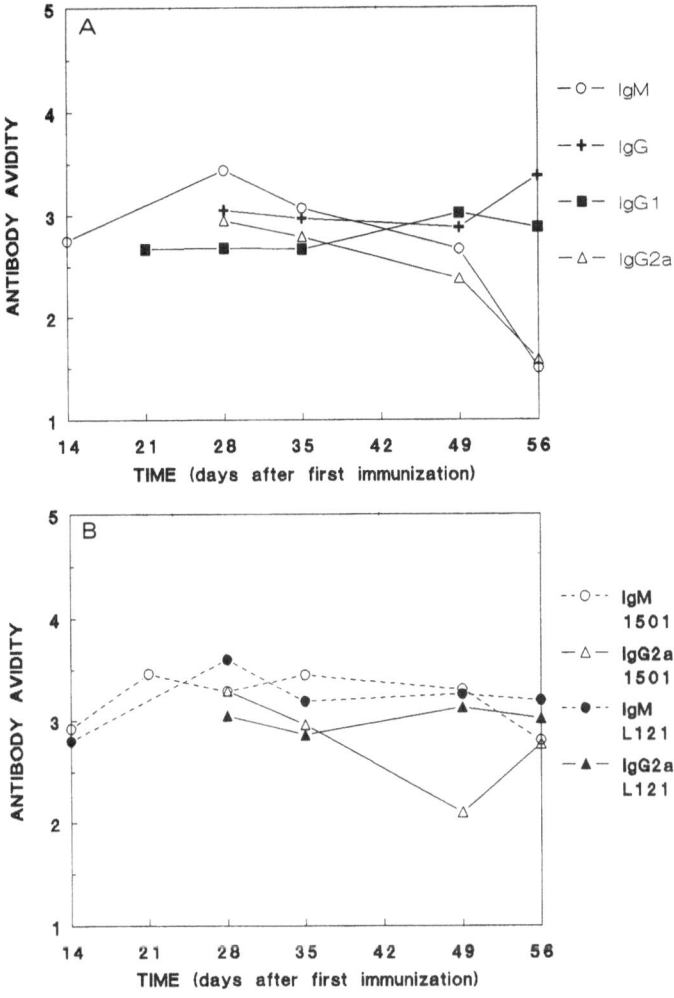

Fig. 7. Serum antibody level and antibody avidity vs time in pheno-
typically Xid mice. Antibody avidity for S3PS of IgM, IgG,
IgG1 and IgG2a in sera collected at different times after
immunization with HS-KLH both with (panel B) and without
(panel A) NBPs. The shown avidity values are the mean of
three independent observations.

after injection of NBP-containing oil-in-water emulsions (Hunter and
Bennett, 1984). In mice, inflammation at injection sites was transient but
some copolymers induced local inflammatory reactions, especially during the
first two weeks. Local tissue reaction was most pronounced when the octablock NBPs were used. Similar results were obtained with guinea pigs which
were vaccinated with oil-in-water emulsions containing BSA as antigen
(unpublished results).

CONCLUSIONS

NBPs are potent adjuvants for the immune response against several
antigens. In general, they stimulate antibody responses considerably.
Different antigens, however, may require different NBPs for optimal stimulation of the immune response. The combination of several NBPs in one
vaccine might be a solution for this problem. This combination can be even
more effective than the individual NBPs alone (Hunter et al, 1989). As was
found already during early research on NBP adjuvant activity, some NBPs
efficiently induce cell mediated immune responses.

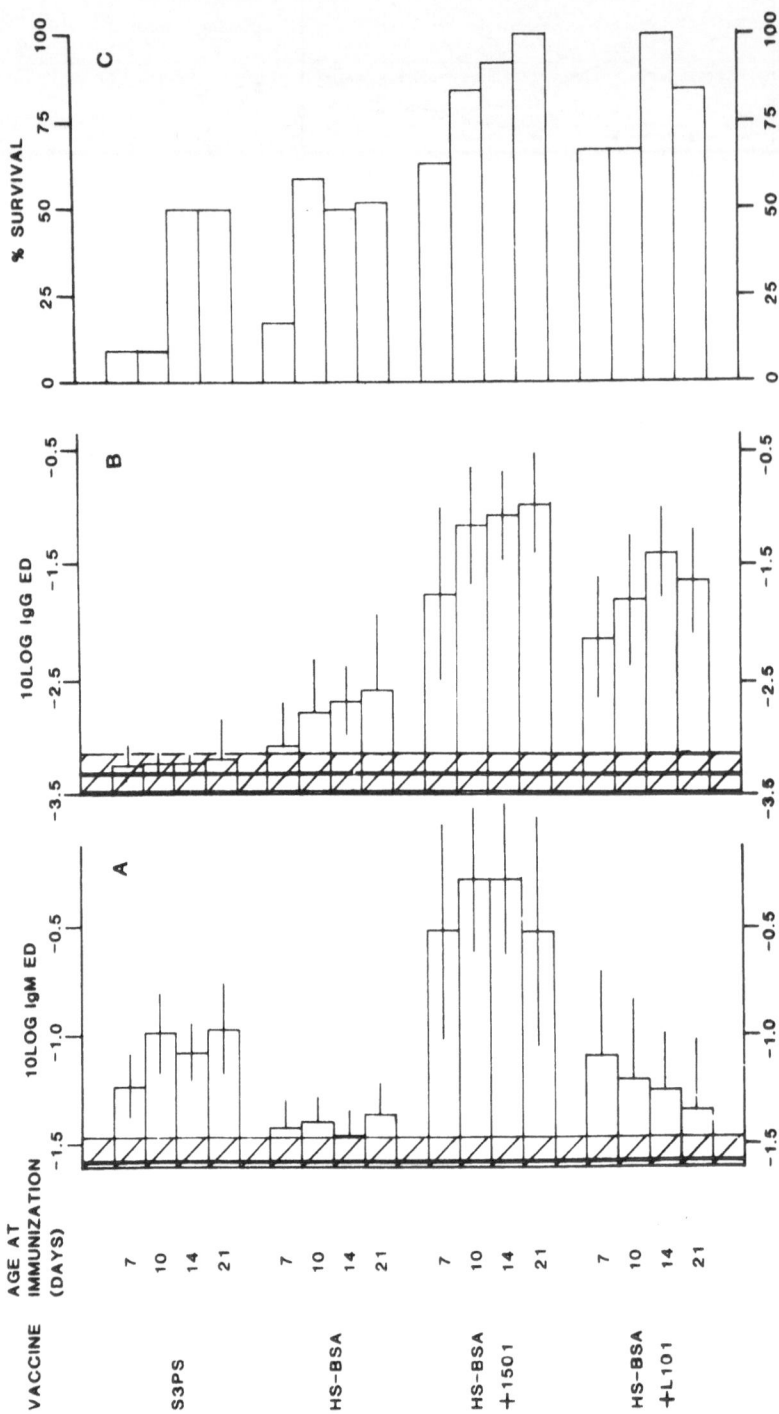

Fig. 8. Induction of circulating IgM and IgG and protection by o/w vaccines in newborn mice. Newborn BALB/c mice (n=12) were immunized i.p. at the indicated age with an o/w emulsion containing 0.5 μg capsular polysaccharide of Streptococcus pneumoniae type 3 or 50 μg HS-BSA and 323 nmol L101 or 1501. IgM (panel A) and IgG (panel B) levels were determined 38 days after immunization. The hatched area on the left of each panel indicates the preimmune antibody level. Protection (panel C) was evaluated 42 days after immunization during a 14 day period following i.p. challenge with 25 x LD_{50} of Streptococcus pneumoniae.

The mode of action of NBPs remains obscure but several mechanisms during the induction stage of the immune response appear to be influenced by NBPs: complement-fixation (Hunter and Bennett, 1984), phagocytosis and Ia-expression by phagocytic cells (Hunter et al, 1989). Other processes during the induction stage (e.g. cell traffic) may also be influenced by NBPs.

Considering the efficacy of NBPs as adjuvants to be incorporated into vaccines, it can be stated that NBPs efficiently stimulate total antibody titers as well as titers of the individual IgG subclasses. This enhanced contribution of several IgG subclasses to the antibody response is parallel-led by an accelerated contribution of several B-cell clones to the antibody response. In general, it can be stated that the immune response is broaden-ed by the use of NBPs. In comparison to CFA, NBPs are more effective in inducing high avidity anti-capsular polysaccharide antibodies (Kenney et al, 1989; Dam et al, 1989a,b; Table 8). These results are obtained with either lattice conjugates or neoglycoproteins. Together with the argument that immune responses in young mice could be enhanced significantly by the use of NBPs, the incorporation of NBPs in vaccines looks promising. When protect-ion experiments were performed, NBPs effectively enhanced survival rates either due to antibody formation, as in the case of S. pneumoniae infection, or due to cell mediated immune responses (Geerlings et al, 1989) as in the case of herpes simplex virus infection.

Before NBPs can be used in practice either in veterinary or in human vaccines, however, more data should become available on occurrence of unwanted side effects, as for example local tissue reactions. Some of the NBP, especially the reversed triblocks, induced local reactions which pre-clude their further use in animals. This is much less the case with others (e.g. L121). The differences between adjuvant doses and toxic doses should therefore be established as soon as possible.

REFERENCES

Allison, A.C. and Byars, N.E., 1986, An adjuvant formulation that selectively elicits the formation of antibodies of protective isotypes and of cell-mediated immunity, J.Immunol.Meth., 95:157.
Allison, A.C. and Byars, N.E., 1987, Vaccine technology: Adjuvants for increased efficacy, Biotechnol., 5:1041.
Behbehani, K., Beller, D.I. and Unanue, E.R., 1985, The effects of beryllium and other adjuvants on Ia expression by macrophages, J.Immunol., 134:2047.
Biozzi, G., Mouton, D., Stiffel, C. and Bouthillier, Y., 1985, A major role of the macrophage in quantitative genetic regulation of immuno-responsiveness and anti-infectious immunity, Adv.Immunol., 36:189.
Byars, N.E. and Allison, A.C., 1987, Adjuvant formulation for use in vaccines to elicit both cell-mediated and humoral immunity, Vaccine, 5:223.
Cahn, A. and Lynn, J.L., Jr., 1983, Surfactants and detersive systems, in: "Encyclopedia of Chemical Technology", H.F. Mark, D.F. Othmer, C.G. Overberger and G.T. Seaborg, eds., Vol. 22, John Wiley and Sons, New York.
Dam, G.J. van, Verheul, A.F.M., Zigterman, G.J.W.J., De Reuver, M.J. and Snippe, H., 1989a, Nonionic block polymer surfactants enhance the avidity of antibodies in polyclonal antisera against Streptococcus pneumoniae type 3 in normal and Xid mice, J.Immunol., 143:3049.
Dam, G.J. van, Verheul, A.F.M., Zigterman, G.J.W.J., De Reuver, M.J. and Snippe, H., 1989b, Estimation of the avidity of antibodies in poly-clonal antisera against Streptococcus pneumoniae type 3 by inhib-ition ELISA, Molecular Immunol., 26:269.

Dick, W.E. and Beurret, M., 1989, Glycoconjugates of bacterial carbohydrate antigens. A survey and consideration of design and preparation factors, in: "Contributions to Microbiology and Immunology", J.M. Cruse and R.E. Lewis, eds., Karger, Basel.

Egwang, T.G. and Befus, A.D., 1984, The role of complement in the induction and regulation of immune responses, Immunol., 51:207.

Geerligs, H.J., Weijer, W.J., Welling, G.W. and Welling-Wester, S., 1989, The influence of different adjuvants on the immune response to a synthetic pepotide comprising amino acid residues 9-21 of herpes simplex virus type 1 glycoprotein D, J.Immunol.Methods, 124:95.

Griffin, W.C., 1979, Emulsions, in: "Encyclopedia of Chemical Technology", Vol. 8., H.F. Mark, D.F. Othmer, C.G. Overberger and G.T. Seaborg, eds., John Wiley and Sons, New York.

Hilgers, L.A.Th., Zigterman, G.J.W.J. and Snippe, H., 1989, Immunomodulating properties of amphiphilic agents, in: "Autoimmunity and Toxicology, Immunedisregulation Induced by Drugs and Chemicals", M.E. Kammuller, N. Bloksma, and W. Seinen, eds., Elsevier Science Publishers.

Howerton, D.A., Hunter, R.L., Ziegler, H.K. and Check, I.J., 1990, Induction of macrophage Ia expression in vivo by a synthetic block copolymer, L 81, J.Immunol., 144:1578.

Hunter, R.L. and Bennett, B., 1984, The adjuvant activity of nonionic block polymer. II. Antibody formation and inflammation related to the structure of triblock and octablock copolymers, J.Immunol., 133:3167.

Hunter, R.L. and Bennett, B., 1986, The adjuvant activity of nonionic block polymer surfactants. III. Characterization of selected biologically active surfaces, Scand.J.Immunol., 23:287.

Hunter, R.L. and Bennett, B., 1987, Modulation of antigen presentation and host mediators by block polymer adjuvants, in: "Progress in Leukocyte Biology", Vol. 6, J.A. Majde, ed., Alan R. Liss, New York.

Hunter, R.L., Strickland, F. and Kezdy, F., 1981, The adjuvant activity of nonionic block polymer surfactants. I. The role of hydrophile-lipophile balance, J.Immunol., 127:1244.

Hunter, R.L., Bennett, B., Howerton, D., Buynitzky, S. and Check, I.J., 1989, Nonionic block copolymer surfactants as immunological adjuvants: Mechanisms of action and novel formulations, in: "Immunological Adjuvants and Vaccines", G. Gregoriadis, A.C. Allison and G. Poste, eds., Plenum Press, New York.

Kenney, J.S., Hughes, B.W., Masada, M.P. and Allison, A.C., 1989, Influence of adjuvants on the quantity, affinity, isotype and epitope specificity of murine antibodies, J.Immunol.Methods, 121:157.

Kido, N., Nakashima, I. and Kato, N., 1984, Correlation between strong adjuvanticity of Klebsiella 03 lipopolysaccharide and its ability to induce interleukin-1 secretion, Cell.Immunol., 85:477.

Klerx, J.P.A.M., van Dijk, H., Kouwenberg, E.A. van der Maaden, W.J. and Willers, J.M.N., 1986, Effects of immunological adjuvants on the mouse complement system. II. Anti-complementary effects of surface-active compounds, Int.J.Immunopharmac., 8:47.

Loos, M. and Bitter-Suermann, D., 1976, Mode of interaction of different polyanions with the first (C1) the second (C2) and the fourth (C4) component of complement. IV. Activation of C1 in serum by polyanions, Immunology, 31:931.

Nakashima, S. and Kamikawa, H., 1984, Accelerated expansion of antibody heterogeneity by complete Freund's adjuvant during the response to bacterial -amylase, Immunology, 53:837.

Pepys, M.B., 1974, Role of complement in induction of antibody production in vivo. Effect of cobra factor and other C3-reactive agents on T-dependent and T-independent antibody responses, J.Exp.Med., 140:126.

Snippe, H., De Reuver, M.J., Strickland, F., Willers, J.M.N. and Hunter, R.L., 1981, Adjuvant effect of nonionic block polymer surfactants in humoral and cellular immunity, Int.Archs.Allergy Appl.Immunol., 65:390.

Svenson, S.B. and Lindberg, A.A., 1979, Coupling of acid labile Salmonella specific oligosaccharides to macromolecular carriers, J.Immunol.Methods, 25:323.

Unanaue, E.R. and Allen, P.M., 1987, The basis for the immunoregulatory role of macrophages and other accessory cells, Science, 236:551.

Verheul, A.F.M., Versteeg, A.A., De Reuver,, M.J., Jansze, M. and Snippe, H., 1989, Modulation of the immune response to pneumococcal Type 14 capsular polysaccharide-protein conjugates by the adjuvant Quil A depends on the properties of the conjugates, Inf.Immunity, 57:1078.

Zigterman, G.J.W.J., 1988, Nonionic block polymer surfactants enhance vaccine efficacy, Thesis, State University Utrecht, Utrecht, The Netherlands.

EFFECTS OF ADDED CYTOKINES ON IMMUNE RESPONSES AND MEMORY

A.W. Heath[1] and J.H.L. Playfair[2]

[1]Paravax Inc., Mountain View, California, USA
[2]Department of Immunology, UCMSM, London, UK

INTRODUCTION

Immunological adjuvants can be defined as substances which are not specific to a particular antigen, but which can enhance the specific immune response to antigens. In recent years a new group of such molecules has been defined and partially characterised. These are non-antigen specific molecules which affect specific immune responses in various ways - the cytokines. The increasing availability of cytokines in purified form, as recombinant proteins, has enabled a much increased understanding of their actions and, among many other things, has revealed their usefulness as adjuvants.

Interleukin 1

The induction of interleukin-1 release has for a long time been considered to be at least a contributing factor in the adjuvant effects of bacterial immunostimulators such as MDP and LPS (Allison, 1983). Demonstration of the adjuvant effect of IL1 itself was performed in part to support this theory (Staruch and Wood, 1983) and, interestingly, prior to the cloning of the molecule. The authors were able to show that a semi-purified IL1 preparation, administered shortly after the antigen, could enhance the memory response of mice to bovine serum albumin. One of the perceived advantages of using a pure cytokine rather than a bacterial product as an adjuvant would be that unwanted side effects might be avoided. Unfortunately, this might not be true in the case of IL1 as its release could itself be the cause of the pyrogenic effects of MDP, indeed one of the cytokine's original names was "endogenous pyrogen". It was this thought that prompted an Italian group to attempt to alter the molecule to take out the pyrogenic portion, hoping to retain adjuvanticity. Nencioni et al (1987) have derived a nona-peptide, representing amino-acids 163-171 of IL1$_\beta$ which lacked pyrogenicity but retained an adjuvant effect on both T-dependent and T-independent antigens. Reed et al (1989a), have also shown an adjuvant effect of IL1 with sheep red blood cells. These results and our own (Heath et al, 1989a and unpublished) showing an adjuvant effect of IL$_\alpha$ with a malaria vaccine, are in contrast to the findings of Boraschi et al (1990), who showed in a comparative study that IL$_\beta$, but not IL$_\alpha$, had an adjuvant effect in vivo.

Vaccines, Edited by G. Gregoriadis *et al.*
Plenum Press, New York, 1991

Interferon gamma

Interferon gamma (IFN_γ) has now been shown to be an effective adjuvant with an impressive list of antigens (see Table 1). The initial work was performed with a Plasmodium yoelii murine malaria vaccine (Playfair and De Souza, 1987). Simply mixing recombinant IFN_γ with the vaccine prior to immunisation led to enhanced protection, secondary antibody and delayed type hypersensitivity responses. IFN_γ appears to be a fairly versatile adjuvant, being effective over a wide dose range, and perhaps surprisingly showing no inhibition of immunogenicity over the wide dose range we examined (10–50,000 units per dose), at least when mixed with the antigen. IFN_γ may have a practical advantage over IL1 or IL2 in that it is effective given in this way, which is obviously the most convenient. Most of the IL1 studies mentioned above relied upon administration of IL1 two hours after the antigen for adjuvanticity to be apparent. The requirements for IL2 adjuvanticity will be discussed below.

Interleukin 2

IL2 is probably the cytokine which has generated most interest as a vaccine adjuvant and we are aware of four separate applications of IL2 to enhance vaccine efficacy. We will cover the first two in this section.

a) "Classical" type adjuvant effect. IL2 has been utilised as an adjuvant to enhance responses to several infectious agents. Administration of IL2 with the vaccine and for the following five days enhanced protection against Haemophilus pleuropneumoniae (Anderson et al, 1987) and rabies virus (Nunberg et al, 1989). IL2 given with the vaccine and for a longer period of 17 days after, similarly enhanced protection against Herpes simplex virus (Weinberg and Merigan, 1988).

b) Overcoming genetic non-responsiveness. IL2 can overcome genetic non-responsiveness to peptide antigens, but only when emulsified in oil with the antigen (Kawamura et al, 1985; Good et al, 1988).

Tumour necrosis factor

Tumour necrosis factor (TNF) shares many properties with IL1, and adjuvanticity is one of them. TNF_α has been shown to enhance the PFC response to sheep erythrocytes (Ghiara et al, 1988), the cytotoxic T cell response to tumour cells and antibody responses to BSA (Talmadge et al, 1988).

HOW DO CYTOKINES WORK?

Interleukin 1

In earlier studies on IL1, adjuvanticity was attributed to indirect effects, such as IL2 or IL4 induction along with the direct effects of the cytokine on T and B cells (Nencioni et al, 1987; Staruch and Wood, 1983). Recently Reed et al (1989a) showed that in their system, looking at the plaque forming cell response to sheep red blood cells, T helper priming was enhanced by IL1 treatment, and this enhancement was independent of IL2 induction, so direct effects may play the more important role.

Interferon γ

We have considered three possible mechanisms of action of IFN_γ as an adjuvant. These included the induction of a second mediator (such as IL1), effects on lymphocyte traffic, and enhancement of antigen presentation via increased MHC class II expression on antigen presenting cells. Our results are described in greater detail elsewhere (Heath et al, 1989b; 1991) but briefly, we have found the following:

i) In transfer experiments with peritoneal cells given shortly after inject- ion of antigen and/or IFN$_\gamma$, the cells which induced the best immune response were those which had contacted both antigen and adjuvant. Mice receiving these cells fared better than those which received a mixture of cells from antigen-injected mice and IFN$_\gamma$-injected mice, suggesting that the release of a second mediator is not of major importance to the adjuvanticity of IFN$_\gamma$.

ii) While IFN$_\gamma$ does induce homing of labelled lymphocytes to an injection site in the skin, this homing was highest in strains of mice in which IFN$_\gamma$ adjuvanticity was lowest, such as BALB/c, and low in strains in which IFN$_\gamma$ was a very effective adjuvant, such as CBA/Ca. On further examination of homing to IFN$_\gamma$-induced in BALB/c mice, we found that homing was predominant- ly or entirely by CD8+ lymphocytes. The relationship of this finding to the lack of adjuvanticity in BALB/c is still uncertain but clearly of interest. Monoclonal antibody depletion of CD8+ cells prior to immunisation did not restore adjuvanticity, although we had previously shown a positive effect of this treatment in another strain of mice (Heath et al, 1989a).

iii) In an examination of the same mouse strains for enhancement of MHC class II expression on peritoneal cells by IFN$_\gamma$, we showed the opposite effect to that found above. That is, those strains in which IFN was an effective adjuvant showed an enhancement of MHC class II expresssion in response to an intraperitoneal injection, whereas BALB/c mice showed little or no response. Again the relationship of this lack of response to the CD8+ cell homing described above has yet to be determined.

In summary, these results would point to enhancement of antigen pres- entation by increased MHC class II expression as probably the major mechan- ism of action of IFN$_\gamma$ as adjuvant.

Interleukin 2

Classical adjuvant effect: The mode of action of multiple doses of IL2 in enhancing protection against several infectious diseases is probably due to expansion of antigen-primed T cell clones. In the work we have mentioned (Anderson et al, 1987; Nunberg et al, 1989) and in our own unpublished work, a single IL2 injection has been ineffective and dosage must continue after antigen exposure, which would support this explanation.

Overcoming genetic non-responsiveness: The administration of IL2 as an emulsion with antigen restored antibody, but not cell-mediated immune res- ponsiveness to the antigen. This effect could not be achieved by the approach above, which seems not to raise antibody levels so the two effects are apparently caused by distinct mechanisms. The authors' explanation for the emulsion effect was that IL2 is acting to enhance the response of small numbers of antigen-specific B cells that might be present in the local lymph node (Good et al, 1988).

IMMUNODEFICIENCY: A SPECIAL APPLICATION?

Any vaccination programme, and particularly one aimed at developing countries, will now have to face the problem of large numbers of subjects being immunodeficient, through AIDS, malnutrition or other infections. Thus there is now a particular need for adjuvants to enhance vaccine efficacy in these groups. All three of the major cytokine adjuvants have been utilised in immunocompromised hosts of various types (see Table 1).

Table 1. Use of Cytokines as Immunological Adjuvants in Immunodeficiency

Immunodeficiency	Cytokine	Immune response	Reference
Biozzi low responder mice	IFN$_\gamma$	Malaria vaccine protection	Heath et al, 1989a
Low affinity antibody mice	IFN$_\gamma$	Malaria vaccine protection	"
CD4+ depleted mice	IFN$_\gamma$	Malaria vaccine protection	"
Ageing mice	IL1	T helper priming	Frasca et al, 1989
Irradiated mice	IL1	T helper priming	"
T cruzi immunosuppression	IL1	Antibody	Reed et al, 1989b
Genetic non-responsiveness	IL2	Antibody response to peptides	Good et al, 1988
Nude mice	Vaccinia-IL2	Probably NK cell mediated	
Nude mice	Vaccinia-IFN$_\gamma$	?	
Dialysis patients	IL2	Antibody to Hepatitis B vaccine	Meuer et al, 1989
Dialysis patients	IFN$_\gamma$	Antibody to Hepatitis B vaccine	

Interleukin 1

ILβ was shown to be effective in restoring T cell function in mice rendered immunodeficient either through ageing or irradiation (Frasca et al, 1989); in addition IL1 could overcome the suppression of antibody responses caused by T cruzi infection in mice (Reed et al, 1989b).

Interferon γ

We have shown IFN$_\gamma$ to be a particularly effective adjuvant in a variety of immunologically abnormal animals. IFN$_\gamma$ was an effective adjuvant along with our murine malaria vaccine in mice genetically predisposed to produce low quantities of antibody (Biozzi low responders), or antibody of low affinity, or mice partially depleted of CD4+ cells by monoclonal antibody treatment prior to immunisation. In the latter group, which we included because of its relevance to AIDS patients, the use of IFN$_\gamma$ as an adjuvant produced a response as good as that seen with IFN$_\gamma$ in normal animals (Heath et al, 1989a), though it did not restore the normal response seen in non CD4-depleted mice. IFN$_\gamma$ was also effective in a group of haemodialysis patients who had responded poorly to a hepatitis B vaccine (Quiroga and Careno, 1989).

80

Interleukin 2

In another clinical study (Meuer et al, 1989) a single dose of IL2 was effective in haemodialysis patients who had previously failed to respond to hepatitis B vaccination. The patients were an unusual group in that their IL2 receptor levels were particularly high, an observation which may help to explain the effectiveness of a single dose of IL2 in this instance.

CYTOKINE GENES AS ADJUVANTS

One of the possible problems weighing against the advantages of vaccinia virus as a live carrier for cloned genes, is that of administration to immunocompromised hosts, which may give rise to a dangerous disseminated infection (Redfield et al, 1987). For this reason, cytokine genes have also been inserted into vaccinia virus in attempts to reduce its pathogenicity. Most of this work has been done with IL2, and insertion of this gene reduces pathogenicity sufficiently for the infection to resolve even in nude mice, in which it is usually fatal (Ramshaw et al, 1987; Flexner et al, 1987). This is thought to be due to enhancement of NK activity in the host). The gene for IFN$_\gamma$ has also been inserted into the virus and results in a similar recovery from infection on inoculation of nude mice (Yilma et al, 1987; Kohonen-Corish et al, 1990).

FURTHER IMPROVEMENTS

There are several possible routes to increasing the adjuvant effects of cytokines described above. Perhaps the most obvious is to use them in combination. As interactions between cytokines are complex, synergism for adjuvanticity will probably have to be determined by experimental assessment of combinations decided upon by informed guesswork or by systematic screening. We have so far had no spectacular success in testing combinations of IFN$_\gamma$ with other cytokines, including IL1 and IL2, at least administered mixed together. But clearly very much more testing needs to be performed, although even to keep pace with the rapid proliferation of interleukins would be a daunting task!

Our work on the mode of action of IFN$_\gamma$ as an adjuvant, described earlier, led to another intriguing possibility for enhancement of adjuvanticity. We found that in order to be an effective adjuvant, IFN$_\gamma$ should contact the same antigen presenting cells as antigen (Heath et al, 1989b). We therefore decided to increase the probability of this occurring by physically coupling the two. By biotinylating IFN$_\gamma$ and utilising avidin as antigen, we were able to enhance the DTH response to avidin in excess of the enhancement provided by mixing IFN with avidin (Heath and Playfair, 1990). Another possibility for the enhancement of cytokine adjuvanticity might be to physically alter the molecule cytokine. This has been described already in the case of IL$_\beta$ as mentioned earlier and more studies of this type will probably follow.

DIRECTION OF THE IMMUNE RESPONSE

On the basis of the examination of murine T helper cell clones, CD4+ T helper cells have been divided into two groups. The TH1 type cells produce IFN$_\gamma$ and IL2 and are responsible for help for cellular responses, such as DTH, and the TH2 type cells produce IL4 and IL5 and provide help for humoral immune responses (reviewed by Mosman and Coffman, 1989). There are cases in which a TH2 type response is thought to be deleterious to protection, perhaps the clearest example being <u>Leishmania</u> infection (Scott et al,

1989), and conversely there are cases when an antibody response without a cellular immune response might be desired. The use of cytokines as immunological adjuvants may provide an opportunity to try and push the immune response in a particular direction.

It has recently been shown that only TH2 cells have receptors for IL1 and respond to this cytokine by proliferation (Lichtmann et al, 1988; Weaver et al, 1988). Thus the use of IL1 as an adjuvant might be expected to enhance the TH2 response selectively, increasing T cell help for antibody responses without enhancing DTH. It is known from the studies we have mentioned above that both antibody responses and T cell help for antibody responses are enhanced, but to our knowledge enhancement of TH1 mediated responses was not assessed in these studies. However, this idea has been indirectly shown to be of possible value; enhancement of the TH2-like response by the use of alum as an adjuvant and of the TH1-like response by the use of complete Freund's adjuvant was shown to be dependent on IL1 induction by the former (Grun and Maurer, 1989). In contrast, the use of IL2 as an adjuvant has been reported to give protection against a number of infectious diseases without a detectable enhancement of antibody responses (Anderson et al, 1987; Nunberg et al, 1989), suggesting that in this case, the TH1 like response may be selectively boosted.

While we have shown enhanced T cell help for antibody and enhanced DTH following the use of IFN$_\gamma$ as an adjuvant, there are two instances when a more directed enhancement of TH1 responses may have been obtained. We have already mentioned one of these, the case of IFN$_\gamma$-antigen conjugation, where we showed a significant enhancement of DTH responses but not of antibody titres following conjugation (Heath and Playfair, 1990). The other was when Scott et al (1990) showed protection of BALB/c mice following Leishmania vaccination with IFN$_\gamma$ and C. parvum as adjuvants. BALB/c mice are normally extremely susceptible to Leishmania infection because they tend to produce immune responses slanted towards the TH2 type. As mentioned above, we were unable to show IFN$_\gamma$ to be a useful adjuvant in BALB/c mice with malaria vaccine. Whether the success of Scott et al (1990) is related to the disease, or to the addition of C. parvum or to another factor has yet to be determined.

Contrasting with what could be a specific TH1 enhancement by IFN$_\gamma$ in the above case, is a very potent suppression of TH2 responses that we have noted when IFN$_\gamma$ is administered two days prior to our vaccine or to another antigen (Heath et al, 1991). We found that while DTH responses were enhanced by this procedure, antibody responses and T cell help for antibody responses were inhibited. This effect was not due to the induction of CD8+ suppressor cells and the possibility that T helper responses were being differentially affected was supported by the finding that IgG1 titres were suppressed, while IgG2a titers were normal or raised; switching to these two subclasses is thought to be positively influenced by IL4 and IFN$_\gamma$ respectively (Snapper and Paul, 1987). Although we are as yet unsure of the mechanism of this down-regulation of the TH2 type response, we were able to achieve some reversal by administering IL1 to the animals just after the antigen. As mentioned above, only the TH2 subset possesses receptors for this cytokine, so our current working hypothesis is that the early administration of IFN$_\gamma$ leads to a defect in antigen presentation to the TH2 subset, which may occur through an abrogation or exhaustion of IL1 production.

While it seems there are already several ways to push the immune response towards TH1 or TH2 type responses, there will surely be many more developments in this field, perhaps involving anti-cytokine antibodies, anti-receptor antibodies, or more cytokines. The in vivo effects of IL-10, the TH2-produced cytokine synthesis inhibitory factor for TH1 cells, will be interesting to observe and this molecule has recently been cloned (Moore et al, 1990).

   In conclusion, while unwanted immune responses to new vaccines might be controlled by the addition of desirable epitopes and the elimination of undesirable ones, the selecting of the desired cytokine adjuvants may provide a further level of control.

REFERENCES

Allison, A.C., 1983, Immunological adjuvants and their mode of action, in: "New Approaches to Vaccine Development", R. Bell and G. Torrigiani, eds., WHO, Schwabe and Co., AG., Basel.

Anderson, G., Urbana, O., Fedorka-Cray, Newell, A., Nunberg, J. and Doyle, M., 1987, Interleukin 2 and protective immunity in Haemophilus pleuropneumoniae. Preliminary studies, in: "Vaccines '87", Cold Spring Harbor Press.

Boraschi, D., 1990, Differential activity of interleukin 1a and interleukin 1b in the stimulation of immune responses in vivo, Eur.J.Immunol., 20:317.

Flexner, C., Hugin, A. and Moss, B., 1987, Prevention of vaccinia virus infection in immunodeficient mice by vector-directed IL-2 expression, Nature, 330:259.

Frasca, D., Boraschi, D., Baschierl, S., Bossu, P., Tagliabue, A., Adorini, L. and Doria, G., 1989, In vivo restoration of T cell function by human IL-1β or its 16-171 nonapeptide in immunodepressed mice, J.Immunol., 141:2651.

Ghiara, P., Borashi, D., Nencioni, L., Ghezzi, P. and Tagliabue, A., 1988, Enhancement of in vivo immune response by tumour necrosis factor, J.Immunol., 139:3676.

Good, M.F., Pombo, D., Lunde, M.N., Maloy, W.L., Halenbeck, R., Koths, K., Miller, L.H. and Berzofsky, J.A., 1988, Recombinant human IL-2 overcomes genetic nonresponsiveness to malaria sporozoite peptides, J.Immunol., 141:972.

Grun, J.L. and Maurer, P.H., 1989, Different T helper subsets elicited in mice utilising two different adjuvant vehicles; the role of endogenous interleukin in proliferative responses, Cell Immunol., 121:134.

Heath, A.W. and Playfair, J.H.L., 1991, Conjugation of interferon gamma to antigen enhances its adjuvanticity, Immunology, in press.

Heath, A.W., Devey, M.E., Brown, I.N., Richards, C.E. and Playfair, J.H.L., 1989a, Interferon gamma as an adjuvant in immunocompromised mice, Immunology, 67:520.

Heath, A.W., Haque, N.A., De Souza, J.B. and Playfair, J.H.L., 1989b, Interferon gamma as an effective immunological adjuvant, in: "Vaccines '89", Cold Spring Harbor Press.

Heath, A.W., Nyan, O., Richards, C.E. and Playfair, J.H.L., 1991, Effects of interferon gamma and saponin in lymphocyte traffic are inversely related to adjuvanticity and enhanced MHC class II expression, Int.Immunol., 3:285.

Kawamura, H., Rosenberg, S. and Berzofsky, J.A., 1985, Immunization with antigen and interleukin 2 in vivo overcomes Ir gene low responsiveness, J.Exp.Med., 162:381.

Kohonen-Corish, M.R.J., King, N.J.C., Woodhams, C.E. and Ramshaw, I.A., 1990, Immunodeficient mice recover from infection with vaccinia virus expressing interferon gamma, Eur.J.Immunol., 20:157.

Lichtmann, A.H., Chin, J., Schmidt, J.A. and Abbas, A.K., 1988, Role of interleukin 1 in the activation of T lymphocytes, Proc.Nat.Acad.Sci. USA., 85:9699.

Meuer, S.C., Dumann, H., Meyer zum Buschenfelde, K.H. and Kohler, H., 1989, Low-dose interleukin-2 induces systemic immune responses against HBsAg in immunodeficient non-responders to hepatitis B vaccination, Lancet, i:15.

Moore, K.W., Viera, P., Florentino, D.F., Trounstine, M.L., Khan, T.A. and Mosman, T.R., 1990, Homology of cytokine synthesis inhibitory factor (IL-10) to the Epstein-Barr virus gene BCRF1, Science, 248:1230.

Mosman, T.R. and Coffman, R.L., 1989, Different patterns of lymphokine secretion lead to different functional properties, Ann.Rev.Immunol., 7:145.

Nencioni, L., Villa, L., Tagliabue, A., Antoni, G., Presentini, R., Perin, F., Silvestre, A. and Boraschi, D., 1987, In vivo immunostimulating activity of the 163-171 peptide of human IL-1β, J.Immunol., 139:800.

Nunberg, J., Doyle, M.V., York, S.M. and York, C.J., 1989, Interleukin 2 acts as an adjuvant to enhance the potency of inactivated rabies virus vaccine, Proc.Nat.Acad.Sci.USA., 86:4230.

Playfair, J.H.L. and De Souza, J.B., 1987, Recombinant gamma interferon is a potent adjuvant for a malaria vaccine in mice, Clin.Exp.Immunol., 67:5.

Quiroga, J.A. and Carreno, V., 1989, Interferon and hepatitis B vaccine in haemodialysis patients, Lancet, ii:1264.

Ramshaw, I.A., Andrew, M.E., Phillips, S.M.N., Boyle, D.B. and Coupar, B.E.H., 1987, Recovery of immunodeficient mice from a vaccinia virus/IL2 recombinant infection, Nature, 329:545.

Redfield, R.R., Wright, D.C., James, W.D., Jones, T.S., Brown, C. and Burke, J.S., 1987, Disseminated vaccinia in a military recruit with human immunodeficiency disease, New Eng.J.Med., 316:673.

Reed, S.G., Pihl, D.K., Conlon, P.J. and Grabstein, K.H., 1989a, IL1 as adjuvant. Role of T cells in the augmentation of specific antibody production by recombinant human IL1α, J.Immunol., 142:3129.

Reed, S.G., Pihl, D.K., Grabstein, K.H., 1989b, Immune deficiency in chronic Trypanosoma cruzi infection. Recombinant IL1 restores T helper function for antibody production, J.Immunol., 142:2067.

Scott, P., Pearch, E., Cheever, A.W., Coffman, R.L. and Sher, A., 1989, Role of cytokines and CD4+ T cell subsets in the regulation of parasite immunity and disease, Immun.Rev., 112:161.

Snapper, C.M. and Paul, W.E., 1987, Interferon gamma and B cell stimulatory factor-1 reciprocally regulate immunoglobulin isotype production, Science, 236:944.

Staruch, M.J. and Wood, D.D., 1983, The adjuvanticity of interleukin 1 in vivo, J.Immunol., 130:2191.

Talmadge, J.E., Phillips, H., Schneider, M., Rowe, T., Pennington, R., Bowersox, O. and Lenz, B., 1988, Immunomodulatory properties of recombinant murine and human tumour necrosis factor, Cancer Res., 48:544.

Weaver, C.T., Hawrylowicz, C.M. and Unanue, E.R., 1988, T helper cell subsets require the expression of distinct costimulatory signals by antigen presenting cells, Proc.Nat.Acad.Sci.USA., 85:8181.

Weinberg, A. and Merigan, T.C., 1988, Recombinant interleukin 2 as an adjuvant for vaccine-induced protection. Immunization of guinea pigs with Herpes simplex virus subunit vaccines, J.Immunol., 140:294.

Yilma, T., Breeze, R.G., Ristow, S., Gorham, J. and Leib, S.R., 1985, Immune response of cattle and mice to the G glycoprotein of vesicular stomatitis virus, Adv.Exp.Med.Biol., 185:101.

Yilma, T., Anderson, K., Brechling, K. and Moss, B., 1987, Expression of an adjuvant gene (interferon-γ) in infectious vaccinia virus recombinants, in: "Vaccines '87", Cold Spring Harbor Press.

THE ASSESSMENT AND USE OF ADJUVANTS

D.E.S. Stewart-Tull

Department of Microbiology
University of Glasgow
Glasgow G12 8QQ, UK

In 1988 at the previous NATO ASI a number of procedures were proposed for the assessment of adjuvants (Stewart-Tull, 1989). Insufficient time has elapsed to allow a detailed experimental appraisal of the recommendations but researchers have questioned some of the tests. In addition, vaccinologists interested in the practical uses of adjuvants must face the problems of the proposed phasing out of mineral oils.

POTENTIAL MINERAL OIL BAN

In March 1990, the U.K. Ministry of Agriculture, Fisheries and Food (MAFF) informed producers of their intention to introduce revised "Mineral Hydrocarbons in Food Regulations". It will be necessary to demonstrate the absence of mineral hydrocarbon residues in food. There were two immediate problems for vaccine producers (a) the toxicological potential of mineral oils and (b) the possible retention of mineral hydrocarbons in the meat of vaccinated animals at non-permissible levels.

Twenty years ago some concern was expressed about the toxicity of mineral oils (white oils) refined by the conventional oleum treatment, that this caused contamination with H_2SO_4 and other sulphur-containing compounds. The process of catalytic hydrogenation reduced the amount of sulphur contamination. CONCAWE (Oil Companies' European Organization for Environmental & Health Protection) analysed the products of both processes and found they were similar. It was concluded by JECFA (Joint European Community Food Administration) and MAFF that further studies were required on mineral oil composition but no acceptable daily intake (ADI) for the food industry was given. A major oil company made two 90-day feeding studies which compared the products of the two processes. In the first a no-effect level (NOEL) was not established but in the second there was evidence of toxicity. The Committee on Toxicity (COT) reacted to this limited study and concluded that the apparent toxicity of these compounds could be attributed to all mineral oils. However, we showed that mineral oils from different sources varied in composition (Stewart-Tull et al, 1976); one mineral oil which contained short-chain hydrocarbons, lipid solvents, was toxic. In June 1989, the EEC Scientific Committee for Food (SCF) recommended ADI of 5 µg kg^{-1}/day for oleum refined oils and 50 mg kg^{-1}/day for hydrogenated oils.

Vaccines, Edited by G. Gregoriadis *et al.*
Plenum Press, New York, 1991

The recommended ADI of the EEC-SCF is impracticable in terms of the MAFF directive of "no detectable residues in food taken from treated animals"; an impossible task considering the complex range of components in mineral oils.

In June 1990, CONCAWE agreed, at the request of the SCF, to an independent toxicological survey on four mineral oils. These results will be assessed by SCF and JECFA before a decision is reached. Meanwhile, the Federation Europeenne de la Sante Animale (FEDESA) and the National Office of Animal Health Ltd. (NOAH) U.K. are maintaining a close watch on the situation. The manufacturers point out the long period of safe use of white mineral oils in human beings albeit that there is known accumulation in human tissues. Nevertheless, there is no satisfactory evidence of white mineral oil, acute or chronic toxicity in various animal studies. In their opinion they present a minimal human health concern. This matter must be resolved as MAFF have allowed a period of five years for manufacturers to find alternative formulations for existing oil-adjuvanted veterinary vaccines.

TOXICITY OF VACCINE ADJUVANTS

There is a great need to rationalise what one means by toxicity. For instance squalane (2, 6, 10, 15, 19, 23 - hexamethyltetracosane) is listed under "toxicity hazard", in the Sigma-Aldrich Database, as producing moderate irritation in guinea-pig skin at the 100 mg dose after 24 hr. However, under "health hazard acute effects" it states that it may be harmful by inhalation, ingestion or skin absorption - yet this has been used successfully in some experimental vaccines.

$$\left[CH_3-\overset{\overset{\displaystyle CH_3}{|}}{CH}-CH_2-CH_2-CH_2-\overset{\overset{\displaystyle CH_3}{|}}{CH}-CH_2-CH_2-CH_2-\overset{\overset{\displaystyle CH_3}{|}}{CH}-CH_2-CH_2- \right]_2$$

2,6,10,15,19,23-Hexamethyltetracosane (Squalane)

There were some doubts raised in the ASI 1990 on Vaccines about the use of the creatine phosphokinase test. This may be difficult to do with the mouse and there is a need to accumulate data with the guinea-pig to establish normal levels. In my own experiments the values tended to vary with the age and weight of the guinea-pig. It was suggested that during the next two years the mouse weight gain test over 7 days should be re-examined. Dr. P. Fuchs (Israel) mentioned that some of the agencies required an intra-peritoneal injection of the product. In an alternative approach proposed by Gerard Trouve (France) a rat paw oedema test is favoured; after injection of 0.05 ml of adjuvant into the hind paw the increase in the volume of the paw is measured for 48 hr. No consensus was reached on toxicity testing and it was agreed to bring comparative results to the next meeting.

PYROGENICITY TEST

Some problems have been experienced by workers using the Limulus lysate assay, mainly due to the false positive reactions which can occur with, for example, cell-wall peptidoglycan. It would be beneficial to use a standard LPS preparation (WHO, US Pharmacopoeia). There is a commercial preparation of Salmonella abortus equi LPS, protein-free preparation in the uniform sodium salt form, available as NovoPyrexal (Pyrotell Diagnostik, Germany). There was agreement that the LAL test, if reliable, would be preferable to the rabbit pyrogenicity test but it was agreed that parallel tests should be made between the two. Richard Hjorth (USA) mentioned an alternative test marketed by M.A. Bioproducts, a protein gel type assay. Dr. A. Kiderlen (Germany) said that the LAL test is sensitive to serum components. It appeared that some false positive reactions may be obtained by both of these latter procedures.

STANDARD ADJUVANTS

Alhydrogel and Freund's Complete Adjuvant were maintained in the status of the positive control adjuvants. However, there is international resistance to the continued routine use of FCA. My belief is that FCA has been used so often as the positive control in many experimental tests that it should be possible to record the expected results as a set of standard values. In any case FCA should not be used for the routine production of antisera.

The Veterinary Immunology Committee has set up an Adjuvant Group under the chairmanship of Dr. Bruce Wilkie, University of Guelph, Ontario, Canada. The aim is to make several independent experimental comparisons of a range of adjuvants which could be used as alternatives to FCA; hopefully some advance will be made before the next meeting.

ANIMALS

At the NATO ASI in 1988, several researchers favoured the continued use of mice, although at the time the consensus was to use guinea-pigs. Subsequently, a number of people expressed their concern about the exclusion of mice and the matter was raised again. In 1975, Manclark et al stated that "too often the choice of the test animal is determined by availability, rather than its suitability for the testing procedure. Some attention is given to the sex, age and species of the test animal, but little or no care is given to the selection of animals of uniform genetic composition with predictable test characteristics. Animals are usually randomly bred, sometimes inbred, rarely syngeneic and never selectively bred for the assay procedure." This is obviously something which requires attention.

The use of outbred mice cold introduce a considerable degree of laboratory to laboratory variation. Joel Goodman (USA) pointed out that the genetic background besides the H-2 haplotype is also important, but Richard Hjorth suggested that out of three haplotypes at least one must be reactive. A standard antigen recommended in 1988 was influenza haemagglutinin (Stewart-Tull, 1989), however, Brian Thomas (UK) stressed that guinea-pigs and rabbits do not respond to influenza - the mouse was reinstated! Harm Snippe (Netherlands) suggested that goats might be considered, and the point was enforced by Richard Hjorth because in goats a human dose can be administered; this is not possible in small rodents. Gary van Nest (USA) has worked with goats, baboons, chimpanzees, guinea-pigs and mice; the results from the latter two species were less dependable. Nevertheless, the purpose of these recommendations is to find a suitable system for the comparison of adjuvants. It was agreed that the guinea-pig and inbred mice were suitable. The use of larger animals could be more suitable for testing adjuvanted vaccines before field trials, although Jonas Salk (USA) asked "if you can document safety, why not test efficacy in humans?"

It was agreed to examine the adjuvant effects in inbred strains of mice to determine the possible differences with mice of variable genetic background and similar H-2 haplotypes or with constant genetic background and variable H-2. As shown in Table I, this would be possible with tests in

Table I. Mouse Strains, with either Similar Genetic Background or
          Histocompatability Specificity, Suitable for Adjuvant Testing

	Mouse strain	Histocompatability specificity
1.	BALB/c	$H\text{-}2^d$
2.	DBA/c	$H\text{-}2^d$
3.	C57BL/10	$H\text{-}2b$
4.	C3H	$H\text{-}2^k$
5.	C3H.B10	$H\text{-}2b$

Strains 1 and 2 or 3 and 5 have variable genetic background
but similar histocompatibility specificity. Strains 4 and 5
have similar genetic background but variable histocompatability
specificity. Test either with groups 1, 2, 4 and 5 of mice, or
with groups 3, 4 and 5 of mice.

three or four strains of mice. The strains listed were recommended by
Chella David (USA). It is imperative that mice should be screened for
organisms and viruses which may interfere with experimental results (e.g.
Sendai virus and mouse hepatitis virus (MHV) are known to depress the immune
response).

Pittman (1967) described the correlation between the immunization of
mouse strains to pertussis vaccines and diphtheria and tetanus toxoids and
the associated histamine sensitization. Together with colleagues (Hansen et
al, 1973; Manclark et al, 1975) breeding programmes were developed to maintain genetic stability in the mouse strain. For inbred strains the genetic
base, from which replacement breeders were chosen, was restricted and there
was continued production breeding from litter mates. The first stage is the
formation of a foundation colony in which mating is brother x sister (BxS)
and the pedigrees are controlled and recorded. The sub-lines must have a
common ancestor within four generations to remain genetically pure. The
colony is expanded by further BxS matings to yield the pedigreed expansion
colony which supports the main expansion colony. Random matings from the
expansion colony provide the source of animals for the production colony.
These animals are only randomly bred within four generations to remain
completely inbred. (Dr. Margaret Pittman, personal communication).

MEASUREMENT OF ISOTYPE SPECIFICITY BY ELISA

The interpretation of results of enzyme-linked immunosorbent assays
done in different laboratories can be extremely difficult to compare. It is
essential to state the various parameters of the experimental protocol: the
type and manufacturer of the plate, the amount of antigen added, the antiserum and conjugate dilutions, the substrate, enzyme reaction time, absorbance wavelength and the multititre plate reactor must be stated.

The Microtitre plate

Experiments in my own laboratory and those of two colleagues in the UK
have revealed that the choice of plate may vary with the antigen used. It
is not the purpose here to advertise the product of one manufacturer but
suffice to record that in each case optimum reactions were obtained in each
laboratory with different plates. To make comparisons of adjuvants it is
important to state the type and manufacturer of the plate.

Addition of reagents

In many instances automatic pipettes are not regularly calibrated and the volumes delivered into the microtitre may well vary. It is suggested that the test should be adjusted such that the volumes decrease from antigen - antiserum 1 - antiserum conjugate - substrate. This means that the possibility of exposing any of the reagents to unblocked plastic is removed. Karen Lovgren (Denmark) recommended the use of 0.5% gelatin in washing buffer as the blocking solution. The wells are filled and the plate is left at RT for 1 h. This step is carried out after the reaction with antigen in a Direct ELISA and is an alternative to the use of variable reagent volumes. There may be some denaturation of the antigen after it is bound to the plate and it may be unable to interact with the antibody. This was found with linear, synthetic peptides during the isolation of some monoclonal antibodies.

Development time

As the final development is based on an enzyme-substrate reaction it is important to standardise the time allowed for the experiment to proceed before stopping the reaction. For the purposes of adjuvant comparison it is suggested that a period of 30 min should be used.

Interpretation of results

A brief review of any leading journal reveals the different approaches used to analyse endpoint ELISA results. The time for the development of the chromophore has been mentioned already but the time should be recorded when the enzyme-substrate reaction is quenched with the stopping reagent. Similarly, the time between quenching and reading the absorbance values may be critical because further intensity of the chromophore can occur after quenching.

In a typical ELISA a sigmoid curve is obtained with high-titre antisera. It is often apparent that titres are recorded at the extreme ends of the sigmoid curve. This is most unreliable. In my own experiments, a baseline of 0.2 at A492nm normally avoids this problem. The endpoint should be determined in a semi-quantitative manner from the linear part of the reaction curve.

Some workers use a range of dilutions and determine the 50% endpoint as the antiserum titre. The use of one antiserum dilution makes comparison with the sera derived from other groups of animals or the results of other workers doubtful. The chosen serum dilution may be 1/50 but it may be 1/1000; so the relative values for these may vary considerably. De Savigny and Voller (1980) examined the fundamental problems in analysing ELISA results and concluded that no single method satisfied all criteria. Bishop et al (1984) estimated anti-rotavirus IgG at a single dilution by reference to a standard curve. However, universal standardization would require the distribution of a standard positive antiserum. If standard antisera against the standard adjuvant antigens could be provided for all laboratories the standard curve method would be desirable.

There was general agreement that multiple antiserum dilutions should be used in ELISAs. In our laboratory, the antibody titre is calculated 1) by comparison with the standard curve of a positive serum or 2) by the calculation of theoretical antibody units ml^{-1} serum (Alastair Wardlaw, personal communication).

The variety of ELISA result presentations is considerable and at the meeting no definitive conclusions were reached but the following examples

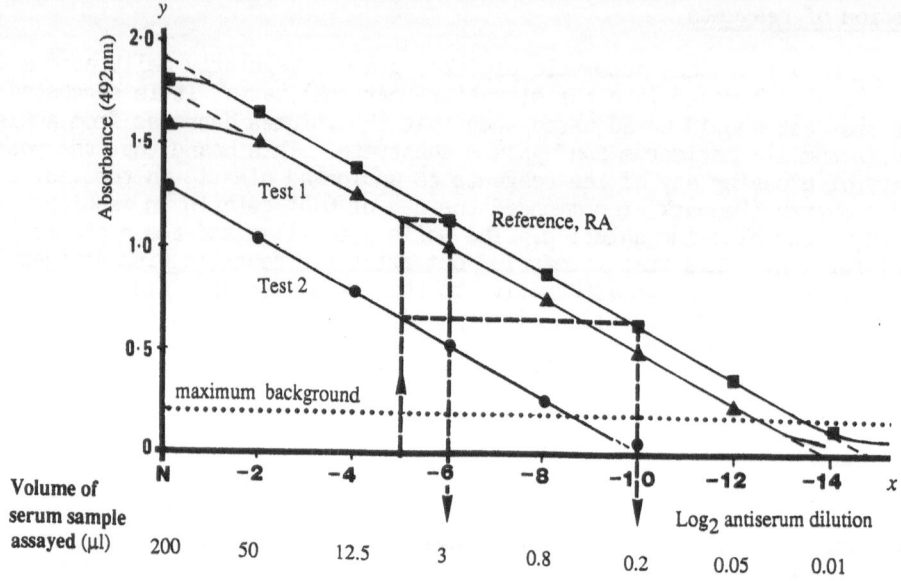

Fig. 1. By assigning an arbitrary unitage of 100 AbU ml^{-1} to the reference antiserum, RA, therefore: AbU ml^{-1} 100 25 6.25 1.5 0.4 0.1 0.025 0.005.

By comparing linear portions of curves and interpolating from test to reference antiserum, RA, as shown above, Test 1 at a dilution of 1/32 (log$_2$ -5.0) contains 1.5 AbU ml^{-1} serum, ie. 48 AbU ml^{-1} neat mouse serum. Test 2, at the same dilution contains 0.1 AbU ml^{-1} serum, ie. 3.2 AbU ml^{-1} neat mouse serum.

(Figs. 1 and 2) may assist adjuvant workers to achieve a measure of uniformity. In addition, there was some support for the final assessment to be based on the calculation of ELISA units (absorbance value x reciprocal of the antiserum dilution). These can be presented in the form of a histogram. There is merit in this method if there is a significant level of non-parallelism between the standard reference and test curves. The parallel line assays shown in the examples below would require such test data from non-parallel curves to be discarded; antibody titres cannot be reliably determined as indicated with II. However, it is still not possible to determine an ELISA titre by this method and one is obliged to compared samples at one serum dilution. It might be more advantageous to take the endpoint at A492nm = 0.5, as indicated below.

ELISA FOR ANTI-INFLUENZA HAEMAGGLUTIN ANTIBODY

As recommended by David Katz, Israel, polystyrene plates are coated with 50 µl of 2 µg ml^{-1} of HA in NaHCO$_3$, pH 9.6, overnight at 4°C. Wash 3 times in ELISA wash buffer (EWB) [EWB = PBS + 0.1% BSA + Tween 20 + merthiolate]. Add 50 µl of appropriate dilutions of test sera, diluted in EWB, in duplicate or triplicate for each sample (1/3 or 1/2 dilution series). Incubate for 2 hours at room temperature. Wash 3 times with EWB. Add 50 µl of peroxidase-labelled goat anti-mouse Ig(G + A + M) (diluted 1/3000). Incubate for 2 hours at room temperature. Wash 3 times with EWB. Add 50 µl of OPD dissolved in pH 4.5 citrate (OPD tablets from Xymed, California; 1 pill/12ml). Incubate for 10 minutes at room temperature. Add 50 µl of 5% (w/v) sodium dodecylsulphate. Read absorbance at 450 nm. Calculate the serum dilution which results in an A$_{450nm}$ of 0.5 (use a computer programme such as "Immunofit", or plot on graph paper). This absorbance value is

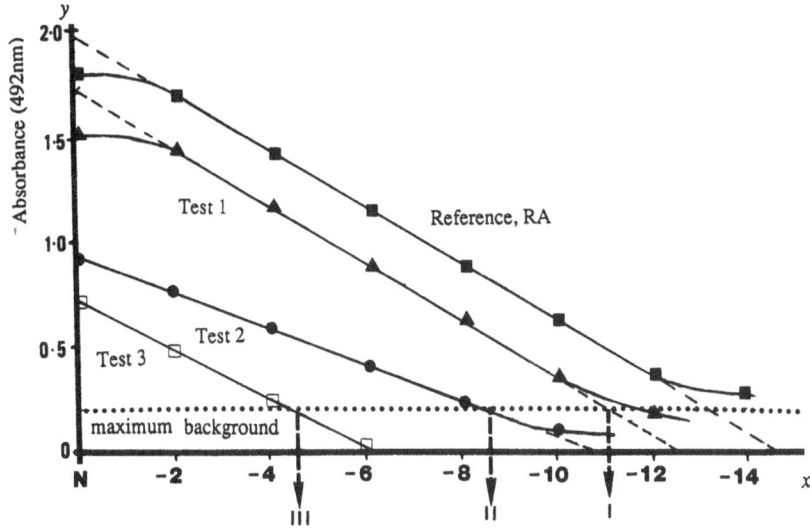

Fig. 2. The linear portions of curves are compared by calculating
test/reference ratios for absorbance (y axis) and $\log_2$ antiserum dilution (x axis) as follows:

y   x

If ratio of test/reference is >0.5 >0.5 then high titre
reading is valid.
If ratio of test/reference is >0.5 <0.5 then low titre
reading is valid.
If ratio of test/reference is <0.5 >0.5 titre is invalid
state Y/X ratio
If ratio of test/reference is <0.5 <0.5 then low titre
reading is valid
In the above example,

y   x

t_1/R =   0.88   0.86 high titre (I) is valid
t_2/R =   0.48   0.73 titre (II) is invalid (Y/X value)
t_3/R =   0.35   0.41 low titre (III) is valid

Antibody titre is the "reciprocal antilog of the $\log_2$ antiserum dilution", read at the intercept point on the x axis.

MEASUREMENT OF LYMPHOKINES

In the published recommendations interleukin-2 (IL-2) was selected for
assay. There was a suggestion that IL-2 did not specifically monitor a
cell-mediated response and a better choice would be IL-3. However, Robert
Bomford (UK) pointed out that T_{H1} and T_{H2} cells both induce IL-3 and we
should monitor the interferon$_\gamma$ (IFN-γ). Tony Allison (USA) mentioned that
the main point was that we needed to be sure that T-cell reactivity was
actually involved. No consensus agreement was reached and it was thought
that results from different laboratories on lymphokine release could be
compared at the next meeting.

REFERENCES

Bishop, R.F., Cipriani, E., Lund, J.S., Barnes, G.L. and Hosking, C.S., 1984, Estimation of rotavirus immunoglobulin G antibodies in human serum samples by enzyme-linked immunosorbent assay: Expression of results as units derived from a standard curve, J.Clin.Microbiol., 19(4):447.

Hansen, C.T., Manclark, C.R., Pittman, M. and Hinkle, W.J., 1973, Selective breeding for HSF sensitivity in the mouse, in "The Laboratory Animal in Drug Testing", Fifth International Committee on Laboratory Animals Symposium, Hannover, 1972, A. Spiegel, ed., Gustav Fischer Verlag, Stuttgart.

Manclark, C.R., Hansen, C.T., Treadwell, P.E. and Pittman, M., 1975, Selective breeding to establish a standard mouse for pertussis vaccine bioassay. II. Bioresponses of mice susceptible and resistant to sensitization by pertussis vaccine HSF, J.biol.Stand., 3:353.

Pittman, M., 1967, Mouse strain variation in response to pertussis vaccine and tetanus toxoid, in, "International Symposium on Laboratory Animals", London, 1966. Symposia Series in Immunological Standardization, Vol. 5, S. Karger, Basel, Munchen and New York.

de Savigny, D. and Voller, A., 1980, The communication of ELISA data from laboratory to clinician, J.Immunoassay, 1:105.

Stewart-Tull, D.E.S., Shimono, T., Kotani, S. and Knights, B.A., 1976, Immunosuppressive effect in mycobacterial adjuvant emulsions of mineral oils containing low molecular weight hydrocarbons, Int.Archs.Allergy appl.Immun., 52:118.

Stewart-Tull, D.E.S., 1989, Recommendations for the assessment of adjuvants (immunopotentiators), in, "Immunological Adjuvants and Vaccines", G. Gregoriadis, A.C. Allison and G. Poste, eds., Plenum Publishing Corporation.

EFFICIENT ANTI-IDIOTYPIC IMMUNIZATION WITH HOMOLOGOUS, VIRUS NEUTRALIZING
MONOCLONAL ANTIBODIES CONJUGATED WITH KLH AND COMBINED WITH QUIL A

Tom A.M. Oosterlaken, Theo Harmsen, Cornelis A. Kraaijeveld
and Harm Snippe

Eykman-Winkler Laboratory of Medical Microbiology
Utrecht University
Utrecht, The Netherlands

Recently, we described an enzyme immunoassay (EIA) on cell cultures for the rapid demonstration of Semliki Forest virus (SFV) neutralizing antibodies in serum (Tiel et al, 1986). In that assay, the spike proteins of SFV, either E_1 or E_2, can be detected in infected L cell monolayers with a horse radish peroxidase (HRPO-) labeled anti-E_1 or anti-E_2 specific monoclonal antibody (MA). Preincubation of the virus inoculum with SFV neutralizing serum reduces virus multiplication and thereby the appearance of the spike proteins on the surface of the infected L cells which are detected by inhibition of absorbance in the EIA. In Fig. 1, the strong neutralizing capacity of MA UM 5.1 (IgG2a) is demonstrated by this assay. The EIA is very useful for the determination of homologous anti-idiotypic activity in sera of mice. Preincubation of a critically neutralizing dose of MA, e.g. $^{-10}$log dilution 5.5 of MA UM 5.1 (Fig. 1), with anti-idiotypic immune serum leads to a diminished capacity of this MA to neutralize and thereby to a rise in absorbance value in the EIA. The anti-idiotypic response was provoked in (female) BALB/c mice. The animals were injected intra/subcutaneously with a mixture of the saponin Quil A and protein A Sepharose-purified MA conjugated with keyhole limpet haemocyanin (KLH). In this study mice were immunized either with purified MA UM 5.1 alone or with a pool of equal amounts of four SFV-neutralizing MAs (UM 1.2, UM 1.4, UM 1.13,and UM 5.1; fifty percent plaque reduction titres respectively: 10^6, $5*10^4$, 10^6 and 10^6). After one (Fig. 3) or two (Fig. 2) intra/subcutaneous booster injections, blood was taken from ether-anaesthetized mice by retro-orbital puncture. In Fig. 2, three individual sera, obtained from mice immunized with MA UM 5.1 alone, were titrated against the idiotype of MA UM 5.1. All mouse sera abrogated the neutralizing capacity of MA UM 5.1 ($^{-10}$log dilution 4.5) up to a dilution of 1/256. The titres of the anti-idiotypic sera, arbitrarily defined as the dilution of serum causing 50% inhibition of the idiotypic MA UM 5.1, were all around 1/512 (Fig. 2). None of the anti-idiotypic sera caused inhibition in the EIA when other SFV-neutralizing MA UM 1.2, UM 1.4 or UM 1.13 were used (results not shown). Normal mouse serum had no effect on the virus neutralizing activity of UM 5.1.

In Fig. 3, the appearance of anti-idiotypic antibodies in serum against MAs UM 1.2 and UM 5.1 is shown after the first subcutaneous booster injection with a pool of four MAs (respectively UM 1.2, UM 1.4, UM 1.13 and UM 5.1, 10 µg each). Only slight quantities of anti-idiotypic antibodies are present before booster injection (day 0), but after that injection anti-

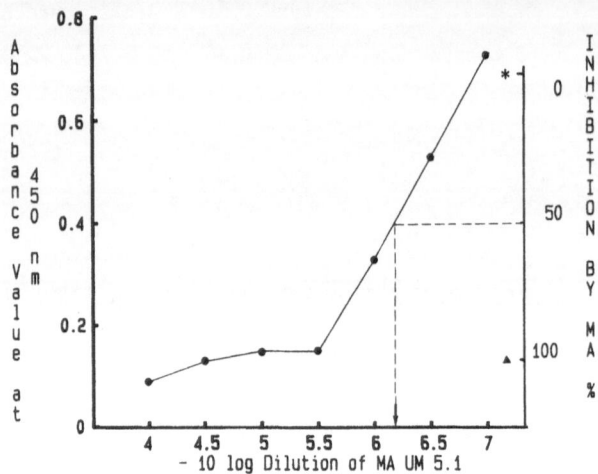

Fig. 1. Determination of the neutralizing capacity of MA UM 5.1 by
         enzyme immunoassay. MA UM 5.1 containing ascitic fluid was
         serially diluted in Dulbecco's minimal essential medium (DMEM),
         supplemented with 5% Calf serum, in wells of 96-well plates
         (catalog no. 3596, Costar, Cambridge, Mass., USA). To the
         dilutions of MA (0.025 ml) a standard dose of 200,000 plaque
         forming units (pfu) of the prototype strain of SFV (Garoff
         et al, 1980) in 0.025 ml complete medium was added and then
         mixed. The virus/MA mixtures were incubated for 1 h at 37°C
         before 20,000 L cells (0.1 ml) were seeded into each well to
         form monolayers. After a multiplication period of 6 h at 37°C
         the monolayers were washed with tapwater, rinsed with phosphate
         buffered saline (PBS) of pH 7.2 and then fixed with 0.05%
         glutaraldehyde. After washing and rinsing with PBS the EIA
         was performed with 1/30,000 diluted conjugate of horseradish
         peroxidase (HRPO) and MA UM 1.4 (IgG2a) containing ascitic
         fluid as described in a previous paper (Tiel et al, 1986).
         The absorbance values were measured with a Titertek Multiscan
         photometer (Flow Laboratories, Irvine, Scotland, UK). Symbols:
         •, single absorbance values measured at 450 nm at different
         -10 log dilutions (4.0, 4.5, 5.0, 5.5, 6.0, 6.5 and 7.0) of MA;
         ∗, mean (n=2) absorbance value of virus control without MA;
         ▲, mean (n=2) absorbance value measured against noninfected
         L cell monolayers. The fifty percent inhibition titre of MA
         in the EIA can be calculated e.g. by graphic interpolation as
         indicated by the arrow in the figure.

idiotypes appear quickly, reaching a maximum level at day 6, and keep constant thereafter until at least day 15. In this particular experiment the
anti-idiotypic antibodies are still detectable at a serum dilution of
1/1000. The results indicate that an immunization procedure with Quil A as
adjuvant is very efficient in inducing high-titred anti-idiotypic sera. The
success of the immunization may be related to the intra/subcutaneous route
of primary and secondary immunization. Other authors found that presentation of antibody to dendritic cells favours an anti-idiotypic immune
response (Urbain et al, 1985). Antigen presenting dendritic cells are
abundantly present in the skin (Choi and Sauder, 1986; Pappaioanou et al,
1986).

    The determination of inhibition of neutralizing activity of an MA by
anti-idiotypic antibodies is a real advantage in comparison to solid phase
immunoassays. Other authors (Gheuens et al, 1981) have described that

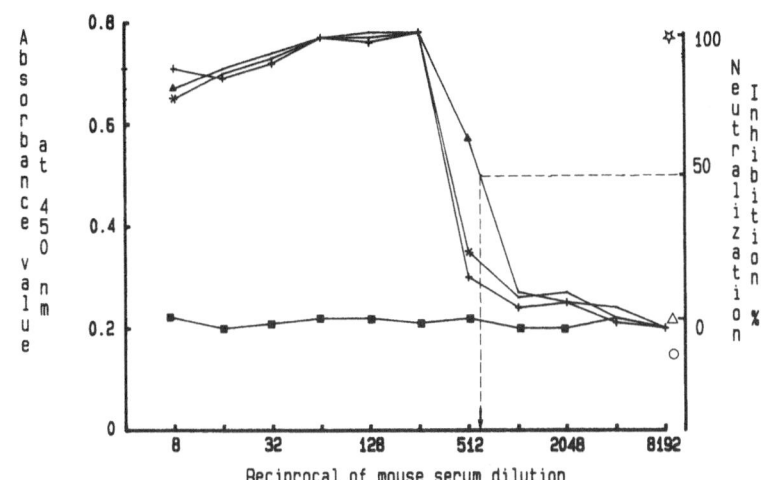

Fig. 2. Inhibition of the virus neutralizing activity of MA UM 5.1
(IgG2a) by mouse anti-idiotypic serum. Anti-idiotypic serum
was induced in mice by injection of protein A Sepharose purified MA UM 5.1, coupled to keyhole limpet haemocyanin (KLH,
Calbiochem, La Jolla, CA, USA), mixed with the adjuvant Quil A
(Sigma Chemicals, St. Louis, MO, USA). In brief: 0.8 mg purified MA UM 5.1 in 0.1 ml PBS was mixed with 1 mg (in 0.2 ml
PBS) of KLH and then coupled to each other by the addition of
0.06 ml 2.5% glutaraldehyde. After 5 min incubation at room
temperature the reaction was stopped with 0.06 ml of 0.2 M
glycine. After addition of 0.58 ml aqua dest, the antigen
was dialyzed overnight at 4°C against aqua dest. The dialyzed
KLH-MA conjugate (1 ml) was mixed with 1 ml (1 mg) of Quil.A.
Twelve week old female BALB/c mice (n=5) were injected intra/
subcutaneously with 0.1 ml of the adjuvant-antigen mixture
(equivalent to 40 μg of MA UM 5.1) at day 0, 21 and 42. Ten
days after the last booster injection blood was taken from
ether-anaesthetized mice by retro-orbital puncture. The
titres of anti-idiotypic antibodies in individual sera were
determined with the neutralization inhibition enzyme immunoassay (NI-EIA). Volumes of 0.025 ml of 1/30,000 diluted
MA UM 5.1 containing ascitic fluid were mixed with 0.025 ml
of two-step (1/8 to 1/8192) dilutions of either immune mouse
sera or normal mouse serum (NMS). After 1 h incubation at
37°C, in wells of 96 well plates, to the MA-serum mixtures
200,000 pfu (0.025 ml) of SFV were added and incubated for
1 h at 37°C. Then to each well 20,000 L cells (0.1 ml) were
seeded to form monolayers. Non-neutralized SFV was allowed
to multiply for 6 h at 37°C before the EIA was performed with
HRPO-labeled MA UM 1.4 (see legend to Fig. 1). Symbols: +,
▲, *, absorbance values (n=1) measured at each dilution of three
individual mouse sera; ■, absorbance values measured with NMS.
Single symbols:☆, mean (n=2) absorbance value of virus control;
○, mean (n=2) absorbance value against noninfected L cells;△ ,
mean (n=2) absorbance value (control) measured in the presence
of SFV and MA UM 5.1 alone. The fifty percent NI-EIA titre of
immune sera can be calculated by graphic interpolation as indicated by the arrows in the figure.

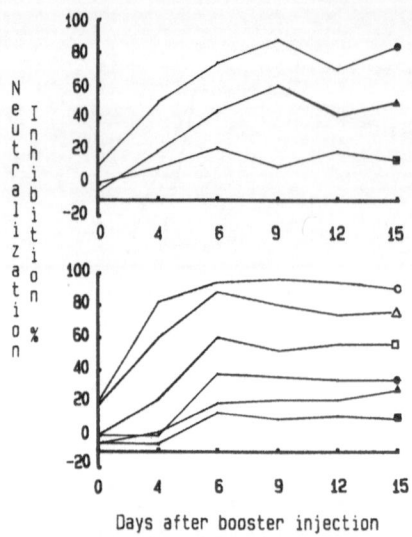

Days after booster injection

Fig. 3. Appearance of anti-idiotypic antibodies in mice after booster injection with pooled MAs. Female BALB/c mice (n=10), 12 weeks of age, were immunized intra/subcutaneously with a pool of purified MAs (UM 1.2, UM 1.4, UM 1.13 and UM 5.1) coupled to KLH and mixed with Quil A (see legend to Fig. 2). Each mouse received an equivalent of about 10 μg of each MA. At day 21, all mice received a booster injection identical to the primary injection at day 0. From one group of 5 mice blood was taken just before (day 0) booster immunization and at day 6 and day 12. Blood from the other group of 5 mice was taken at days 4, 9 and 15. The individual sera were tested for their content of anti-idiotypic antibodies against MA UM 1.2 (Fig. 3a: dilution of MA 1/100,000) and MA UM 5.1 (Fig. 3b: dilution of MA 1/300,000) with the NI-EIA as described in the legend to Fig. 2. The absorbance values given at each time-point and each dilution of anti-idiotypic antibody are means of the single absorbance values obtained with the 5 individual mouse sera. Symbols for 10 log dilutions of anti-idiotypic antibody containing mouse immune sera: ○, 1.0; △, 1.5; □, 2.0; ●, 2.5; ▲, 3.0; and ■, 3.5.

measles hemagglutinin specific MAs are inhibited in their neutralizing activity by homologous anti-idiotypic serum as indicated by diminished neutralization of measles in a plaque reduction test (duration six days). Their immunization protocol, similar to Bona et al (1979) demanded, however, over six immunization instead of our twofold immunization. Another advantage of the SFV model over measles is the possibility to perform protection studies in mice. Previously, we have shown that neutralizing and non-neutralizing MAs protect mice against lethal infection by SFV (Boere et al, 1984; Boere et al, 1985). In preliminary experiments we found that pre-incubation of critically protective doses of MA UM 5.1 with anti-idiotypic serum results in abrogation of protection in vivo (Table I). In this regard it is of interest that anti-idiotypic reagents that recognize idiotypic determinants on MAs to herpes simplex virus type 2, if injected into mice before challenge with herpes virus, cause a shorter survival time in these mice (Kennedy et al, 1984). The authors had no explanation for that observation but it might be due to interference of the idiotypic antibodies with the function of the combining site on early appearing herpes specific IgM molecules.

Table I. Abrogation by Anti-Idiotypic Immune Serum of MA UM 5.1 Mediated Protection Against a Lethal SFV Infection

Treatment of mice*	No survivors/ no. infected**	Mean survival time (days) of non-surviving mice
Phosphate Buffered Saline	0/7	6.1
MA UM 5.1	6/6	
MA UM 5.1 + anti-idiotypic serum against UM 5.1	0/6	6.7
MA UM 5.1 + normal mouse serum	6/6	

*One ml of MA UM 5.1 containing ascitic fluid (dilution 1/10,000) was preincubated for 1 h at 37°C with either 1 ml of diluted (1/25) normal mouse serum or 1 ml of diluted (1/25) anti-idiotypic immune serum against MA UM 5.1. The individual anti-idiotypic sera titrated in a foregoing experiment (Fig. 2) were pooled and used in the present experiment. After pre-incubation, 0.25 ml of each mixture was injected intravenously into groups of mice. Control groups of mice received either the diluent PBS alone or MA UM 5.1 alone.

**Three hours after passive transfer all mice were challenged intra-peritoneally with $10LD_{50}$ of virulent SFV (Bradish et al, 1971) as described previously (Boere et al, 1984). The mice were observed for two weeks after challenge.

The described methods are a major step in the development of anti-idiotypic vaccines and allows further dissection of the idiotypic network (Jerne, 1974).

REFERENCES

Boere, W.A.M., Harmsen, T., Vinje, J., Benaissa-Trouw, B.J., Kraaijeveld, C.A. and Snippe, H., 1984, Identification of distinct antigenic sites on Semliki Forest Virus using monoclonal antiviral activities, J.Virol., 52:575.

Boere, W.A.M., Benaissa-Trouw, B.J., Harmsen, T., Erich, T., Kraaijeveld, C.A. and Snippe, H., 1985, Mechanisms of monoclonal antibody-mediated protection against virulent Semliki Forest Virus, J.Virol., 54:546.

Bona, C., Hooghe, R., Cazenave, D.A., Leguern, C. and Paul, W.E., 1979, Cellular basis of regulation of expression of idiotype. II. Immunity to anti-MOPC-460 idiotype antibodies increases the level of anti-Trinitrophenyl antibodies bearing 460 idiotype, J.exp.Med., 149:815.

Bradish, C.J., Allner, K. and Maber, H.B., 1971, The virulence of original and derived strains of Semliki Forest virus for mice, guinea-pigs and rabbits, J.Gen.Virol., 12:141.

Choi, K.L. and Sauder, D.N., 1986, The role of Langerhans cells and keratinocytes in epidermal immunity, J.of Leukocyte Biol., 39:343.

Garoff, H., Frischauf, A.M., Simons, K., Lehrach, H. and Delius, H., 1980, Nucleotide sequence of cDNA coding for Semliki Forest virus membrane glycoproteins, Nature, 288:236.

Gheuens, J., McFarlin, D.E., Rammohan, K.W. and Bellini, W.J., 1981, Idiotypes and biological activity of murine monoclonal antibodies against the hemagglutinin of measles virus, Infect.Immun., 34:200.

Jerne, N.K., 1974, Towards a network theory of the immune system, Ann.Inst.Pasteur Immunol., 125C:373.

Kennedy, R.C., Adler-Storthz, K., Burns, J.W., Henkel, R.D. and Dreesman, G.R., 1984, Anti-idiotype modulation of Herpes Simplex Virus infection leading to increased pathgenicity, J.Virol., 50:951.

Pappaioanou, M., Fishbein, D.B., Dreesen, D.W., Schwartz, J.K., Campbell, G.H., Summer, J.W. Patchen, L.C. and Brown, W.J., 1986, Antibody response to preexposure human diploid-cell rabies vaccine given currently with chloroquine, New Engl.J.of Med., 314:280.

Tiel, F.H. van, Harmsen, T., Wagenaar, M., Boere, W.A.M., Kraaijeveld, C.A. and Snippe, H., 1986, Rapid determination of neutralizing antibodies in serum by enzyme immunoassay in cell culture using virus specific peroxidase-labeled monoclonal antibodies, J.of Clin. Microbiol., 24:665.

Urbain, J., Brait, M., Bruyns, C., Demeur, C., Dubois, P., Francotte, M., Franssen, J.D., Hiernaux, J., Leo, O., Marvel, J., Meyers, P., Moser, M., Slaoui, M., Tassignon, J., Urbain-Vansanten, G., Van-Acker, A., Wikler, M., Willems, F. and Wuilmart, C., 1985, The idiotypic network: Order from the beginning or order out of chaos? Current Topics in Microbiol. and Immunobiol., 119:127.

PROTEIN CONFORMATION AFFECTS THE EFFICACY OF PERTUSSIS VACCINES

A. Bartoloni, M.G. Pizza, A. Covacci, D. Nucci, L. Nencioni
and R. Rappuoli

Sclavo Research Center
Via Fiorentina 1
53100 Siena, Italy

INTRODUCTION

Production of new and better vaccines to improve man's health is one of
the goals of recombinant DNA technology. In spite of the great number of
papers published on the cloning and expression of genes from a variety of
human and animal pathogens, as yet only the hepatitis B recombinant vaccine
has been successfully introduced for human immunization and there are not
many other promising candidates foreseen in the near future. The general
finding has been that recombinant antigens are immunogenic but induce antibodies which do not recognize, or recognize poorly, the antigens in their
native conformation and therefore they are not effective vaccines. This
finding is particularly true when the natural antigens have a complex
structure as in the case of viral capsids or proteins composed of multiple
subunits.

In our laboratory we experienced similar problems while trying to
develop a new vaccine against pertussis. We report here a study aimed at
understanding the reasons for the lack of efficacy of recombinant pertussis
molecules and how we eventually used recombinant DNA technologies to obtain
molecules which are ideal vaccine candidates.

Whooping cough is a severe disease which each year affects over 60
million children and is responsible for over one million deaths. The vaccine presently available, composed of killed bacteria, is very effective in
preventing disease. However, it is not yet widely used in developing
countries while in Western countries the fear of adverse reactions has
decreased the vaccine acceptance with a resulting increase in morbidity and
mortality due to the disease (Moxon and Rappuoli, 1990). The pressure to
have a new and safer vaccine led to the development of a number of accell-
ular vaccines composed of purified antigens. A vaccine composed of chemically detoxified pertussis toxins (PT) has been shown to be effective in
preventing whooping cough during a large-scale clinical trial carried out in
Sweden (Ad hoc Group, 1988). This showed that PT-based vaccines could be
effectively used, provided that we find better ways to detoxify pertussis
toxin. In fact, the chemical methods used so far either do not completely
inactivate the toxin which can then revert to toxicity (a feature observed
in the vaccines used in Sweden; Storsaeter et al, 1988), or modify the toxin
so heavily that it loses immunogenicity.

Vaccines, Edited by G. Gregoriadis *et al*.
Plenum Press, New York, 1991

The complete and stable detoxification of PT toxin is crucial for the development of safer vaccines because some of the severe side effects associated with pertussis vaccination might be due to trace amounts of active toxin (Steinman et al, 1985).

While a variety of chemical methods to detoxify PT were undertaken by vaccine manufacturers and research laboratories, we decided to tackle the problem from a different point of view and we cloned and sequenced the pertussis toxin operon, with the aim of developing a new vaccine by recombinant DNA technology.

RECOMBINANT PERTUSSIS TOXIN SUBUNITS ARE NOT EFFECTIVE VACCINES

Pertussis toxin is a complex bacterial toxin composed of five subunits named S1 through S5 (Tamura et al, 1982). S1 is an enzyme which intoxicates cells by ADP-ribosylating eukaryotic GTP-binding proteins involved in signal transduction thus modifying the cellular response to exogenous stimuli. Subunits S2, S3, S4 and S5 present in a 1:1:2:1 ratio are involved in binding to receptors on the surface of eukaryotic cells and delivering the toxic S1 across the cell membrane so that it can reach its target.

The genes coding for the five subunits of pertussis toxin were found to be clustered in a fragment of DNA of 3200 base pairs and to be organized as an operon (Nicosia et al, 1986; Locht and Keith, 1986). Since E. coli turned out to be unable to express the entire pertussis toxin operon and assemble the holotoxin, we decided to express separately each individual subunit. Eventually, large amounts of each subunit were expressed by us as fusion proteins in E. coli (Nicosia and Rappuoli, 1987) and as non-fused proteins in E. coli (Burnette et al, 1988a) and B. subtilis (Runeberg-Nyman et al, 1990) by other laboratories. In each case the recombinant S1 subunits were found to maintain their enzymatic activity suggesting that the folding of the recombinant proteins was identical to that of the natural molecule. This suggested that the recombinant molecules could be effective vaccines since monoclonal antibodies raised against the holotoxin but recognizing subunit S1 had been described which were able to neutralize the toxin in vitro and to protect mice against infection with virulent B. pertussis (Sato et al, 1984; Sato et al, 1987).

However, to our surprise, none of the recombinant molecules, including S1, induced toxin neutralizing antibodies or protected mice against infection with virulent bacteria (Nicosia and Rappuoli, 1987). Similar results were obtained with recombinant subunits made by other laboratories. With the aim of understanding why recombinant molecules were unable to induce protection, we characterized the properties of recombinant S1 subunits fusion proteins. As shown in Table I, two lots of partially purified S1 molecules, which had a purity of approximately 50-90%, retained most of the enzymatic activity and were able to induce a high level of antibodies which, diluted up to 1/1000, recognized in a Western blot the natural S1 subunit. However, these antibodies were unable to neutralize the toxin even at a dilution of 1/10. In marked contrast, a rabbit serum raised against the holotoxin which had a similar titer in a Western blot, was able to neutralize the toxin even when diluted 1/1000. These results showed that the natural toxin and the recombinant S1 subunits were inducing two different populations of antibodies and that only those against the natural toxin were protective. To test whether the recombinant S1 molecules had the correct antigenic conformation, we tested them in a radioimmunoassay where they competed with ^{125}I-labelled PT for a monoclonal antibody which we had previously shown to recognize a protective epitope composed of non-contiguous regions of the S1 subunit (Bartolini et al, 1988). Surprisingly, in spite of the presence of the enzymatic activity which had suggested a correct

Table I. Properties of Fused Recombinant S1 Subunits

	Purity	ADP-ribosyl-transferase activity (%)	% of the protein recognized by protective monoclonal antibody	Titer of antibody induced against PT	Titer of neutralizing antibody (CHO cells)
Prep. 1	75%	30	0.03	1/1000	<1/10
Prep. 2	90%	40	0.3	1/1000	<1/10
Native toxin	>98%	100	100	1/1000	1/1000

folding, only 0.3-3% of the recombinant molecules were found to be recognized by the protective monoclonal antibody. This finding suggested the model shown in Fig. 1, where the natural S1 subunit has an enzymatically inactive "closed" conformation which requires the presence of the disulfide, while the recombinant S1 subunit has an enzymatically "open" conformation. As shown in the model, the recombinant S1 subunit would not be able to induce toxin neutralizing antibodies because the protective epitope is formed by three regions which are contiguous in the three-dimensional structure of the natural S1 subunit but not in the recombinant S1 subunit. The model is also confirmed by the following observations: the S1 subunit contains a disulfide bridge which needs to be reduced to obtain enzymatic activity; the regions of the protective epitope are in the proximity of the cysteines which make the disulfide bridge; treatment with strong denaturing agents or boiling is followed by recovery of the enzymatic activity, suggesting that the enzymatically active (open) conformation is energetically favored by the S1 subunit.

Although from our studies we could not conclude whether 0.3-3% of the recombinant molecules are in the closed conformation schematized in Fig. 1 or whether all molecules have a conformation which is only slightly different from the natural conformation and each recombinant molecule is

Fig. 1. Model of the natural S1 subunit (left) and of the recombinant S1 subunit (right). The natural S1 subunit subunit exists in the conformation shown in the left, only in the presence of the B oligomer. Under these conditions three non contiguous regions (black) come together and form a protective epitope. This epitope is not present in the unfolded, enzymatically active recombinant S1 subunit (right).

Table 2. The Effect of Combined Mutations of PT on its Toxicity

Toxic Properties	Combined Mutations: PT-9K/129G PT-13L/129G PT-26L/129G	PT
ADP-ribosyltransferase activity	<0.01%	100%
CHO toxicity	<0.0001%	100%
Lymphocytosis	<0.1%	100%
HSF	<0.1%	100%
Potentiation of anaphylaxis	<0.1%	100%
Affinity for protective monoclonal antibodies	3.4×10^9	3.4×10^9
Presence of T cell epitopes	+++	+++
LD50	Not detectable	0.5 µg
Titer of in vivo protective antibodies induced in guinea pigs	>1/1280	1/1280
ED50	1.7 µg	2.5 µg

recognized by the antibodies with an affinity which is 100 times lower than that of wild type PT, we felt that the model was true enough to explain the experimental data and discourage any further effort to make a vaccine out of recombinant S1 subunit and thus sought alternative ways to develop a vaccine.

CONSTRUCTION OF GENETICALLY DETOXIFIED PERTUSSIS TOXIN MOLECULES

The pioneering work of Uchida et al (1971) had shown that the best way to detoxify a toxin is to modify its gene in order to obtain non toxic molecules which are devoid of enzymatic activity. Therefore, we mutagenized the S1 gene in order to identify aminoacids essential for the enzymatic activity. Three aminoacids, whose substitution abolished the enzymatic activity of the S1 subunit, were identified by our group (Pizza et al, 1988) and several others were described by other groups (Barbieri et al, 1988; Burnette et al, 1988b). These include Arg9, Asn13, Trp26 and Glu129. When the genes containing one of the above-mentioned mutations were introduced in the chromosome of B. pertussis by the method shown in Fig. 2, we obtained strains producing pertussis toxin molecules with a toxicity 10-1000 times lower than wild type toxin. The toxicity could be further reduced to levels lower than 0.0001% by combining mutations 9/129, 13/129 or 26/129 (Pizza et al, 1990; Nencioni et al, 1990). These mutants did not show any of the toxic properties typical of PT such as lymphocytosis (Table II), histamine sensitivity and potentiation of anaphylaxis. In contrast to the wild type

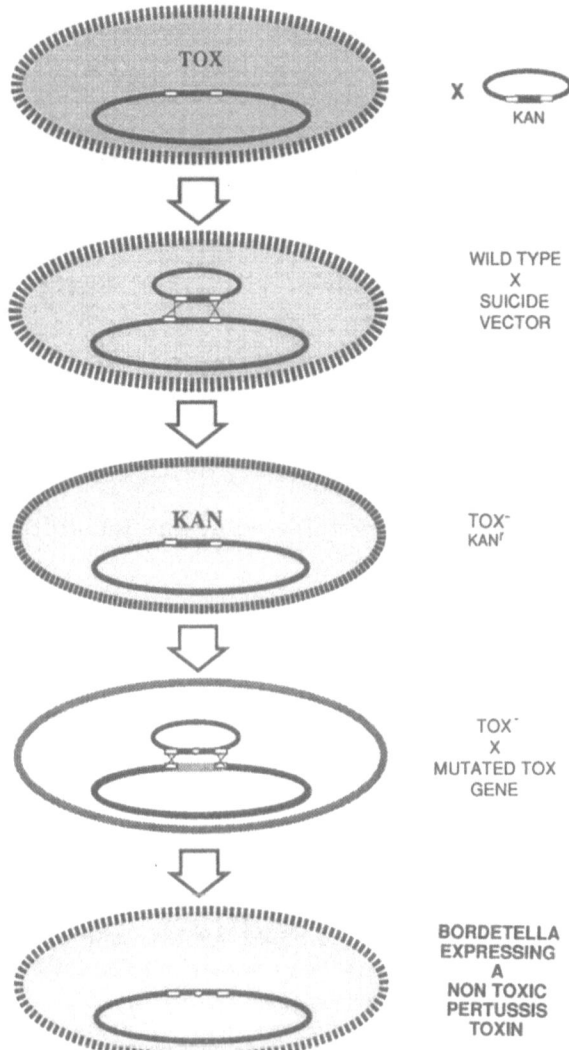

Fig. 2. Schematic representation of the allelic exchange method
used to replace the chromosomal pertussis toxin operon
with the in vitro modified operon.

toxin which is lethal for mice at a dose of 0.1-0.8 µg, up to 50 µg of PT-9K/129G could be injected in mice without detectable toxicity.

The mutants obtained were able to compete with ^{125}I-PT as efficiently as wild type PT for binding to the protective monoclonal antibodies against the S1 subunit and for binding to a polyclonal antibody against the holo-toxin. Similarly, they were recognized by human T cell clones against immunodominant epitopes of the S1 subunit confirming that the mutations introduced had not altered the immunological properties. When tested for the ability to induce toxin neutralizing antibodies, two doses of 3 µg of PT-9K/129G, PT-26L/129G and PT-13L/129G injected i.p. in guinea pigs were found to induce a toxin neutralizing antibody titer of >/1280, a value superior to what is generally obtained with chemically detoxified pertussis toxin. Finally, they were able to protect mice in a dose-dependent fashion,

Table 3. Protective immunity in mice and toxin-neutralizing antibodies from guinea pigs vaccinated with PT-9K/129G

IC CHALLENGE				NEUTRALIZING TITER[a]		
Cellular		Acellular		Acellular		
Dose (ml)	S/T[b]	Dose (μg)	S/T[b]	Dose (μg)	Serum dilution^{-1} 1 dose	2 doses
0.04	12/16	15	13/16	15	80	2560
0.008	5/16	3	9/16	3	40	1280
0.001	3/16	0.6	4/16	0.6	10	320

[a]Serum dilution inhibiting the clustering effect of 120 pg of wild type PT on CHO cells.

[b]Survivors over a total of 16 mice infected by the intracerebral (IC) route with virulent B. pertussis.

from the intracerebral challenge with virulent B. pertussis (Table 3). Since this last test correlates with the efficacy in humans of the cellular vaccine, we believe that the mutants obtained are ideal candidates for vaccine development. Phase I and phase II studies in adult volunteers and in three and 15 month old children confirm the above data.

In conclusion we have shown that while on the one hand it was relatively easy to obtain recombinant molecules with enzymatic activity, on the other hand it was impossible to obtain molcules with the appropriate conformational structure to be used as vaccines. The difficulty of obtaining properly folded recombinant molecules might explain why of the many recombinant enzymes or ligands that have been developed for human use, the hepatitis B vaccine is the only one so far developed.

When the development of recombinant molecules with appropriate conformation is difficult or impossible, we have shown that recombinant DNA techniques can still be of extreme usefulness for the in vitro modification of the gene of interest which can then be reintroduced under appropriate conditions in the natural host for high level expression of the protein of interest.

REFERENCES

Ad hoc Group for the study of pertussis vaccine, 1988, Placebo-controlled trial of two acellular pertussis vaccines in Sweden. Protective efficacy and adverse events, Lancet i:955.
Barbieri, J.T. and Cortina, G., 1988, ADP-ribosyltransferase mutations in the catalytic S-1 subunit of pertussis toxin, Infect.Immun., 56:1934.
Bartolini, A., Pizza, M.G., Bigio, M., Nucci, D., Ashwort, L.A., Irons, L.I., Robinson, A., Burns, D., Manclark, C., Sato, H. and Rappuoli, R., 1988, Mapping of a protective epitope of pertussis toxin by in vitro refolding of recombinant fragments, Biotechnol., 6:709.
Burnette, W.N., Mar, W.L., Ciepalk, W., Morris, C.F., Kaliot, K.T. Marchitto, K.S., Sachdev, R.K., Locth, C. and Keith, J., 1988a,

Direct expression of Bordetella pertussis toxin subunits to high
levels in Escherichia coli, Biotechnol., 6:699.

Burnette, W.N., Ciepalk, W., Mar, V.L., Kaliot, K.T., Sato, H. and Keith,
J.M., 1988b, Pertussis toxin S-1 mutants with reduced enzyme activity
and conserved protective epitope, Science, 242:72.

Locth, C. and Keith, J., 1986, Pertussis toxin gene: nucleotide sequence
and genetic organization, Science, 232:1258.

Moxon, R. and Rappuoli, R., 1990, Modern vaccines: Haemophilus influenzae
infections and whooping cough, Lancet, i:1324.

Nencioni, L., Pizza, M.F., Bugnoli, M., De Magistris, M.T., Di Tommaso, A.,
Giovannoni, F., Manetti, R., Marsili, I., Matteucci, G., Nucci, D.,
Olivieri, R., Pileri, P., Presentini, R., Villa, L., Kreeftemberg,
J.G., Silvestri, S., Tagliabue, A. and Rappuoli, R., 1990,
Characterization of generally inactivated pertussis toxin mutants:
candidates for a new vaccine against whooping cough, Infect.Imm.,
58:1308.

Nicosia, A., Perugini, M., Franzini, C., Casagli, M.C., Borri, M.G.,
Antoni, G., Almoni, M., Neri, P., Ratti, G. and Rappuoli, R., 1986,
Cloning and sequencing of the pertussis toxin genes: operon structure
and gene duplication, Proc.Natl.Acad.Sci.USA, 83:4631.

Nicosia, A. and Rappuoli, R., 1987, Expression and immunological properties
of the five subunits of pertussis toxin, Infect.Imm., 55:963.

Pizza, M., Bartoloni, A., Prugnola, A., Silvestri, S. and Rappuoli, R.,
1988, Subunit S1 of pertussis toxin: mapping of the regions essential
for ADP-ribosyltransferase activity, Proc.Natl.Acad.Sci.USA, 85:7521.

Pizza, M.G., Bugnoli, M., Manetti, R. and Rappuoli, R., 1990, The S1 subunit
is important for pertussis toxin secretion, J.Biol.Chem, in press.

Runeberg-Nyman, K., Olander, R. and Karvonen, M., 1990, Immunogenicity and
protective capacity of pertussis toxin subunits produced in B.
subtilis, in: Vaccines 90. Modern Approaches to New Vaccines,
Including Prevention of AIDS, F. Brown, R.M. Chanock, H.S. Ginsberg,
and R.A. Lerner, eds., Cold Spring Harbor: Cold Spring Harbor
Laboratory.

Sato, H., Ito, A., Chiba, J. and Sato, Y., 1984, Monoclonal antibody against
pertussis toxin: effect on toxin activity and pertussis infections,
Infect.Immun., 46:422.

Sato, H., Sato, Y., Ito, A. and Ohishi, I., 1987, Effect of monoclonal antibody to pertussis toxin on toxin activity, Infect.Immun., 55:909.

Steinman, L., Weiss, A., Adelman, N., Lim, M., Zuniga, R., Ochlert, J.,
Hewlett, E. and Falkow, S., 1985, Pertussis toxin is required for
pertussis vaccine encephalopathy, Proc.Natl.Acad.Sci.USA, 82:8733.

Storsaeter, J., Olin, P., Renemar, B., Lagergard, T., Norberg, R., Romanus,
V. and Tiru, M., 1988, Mortality and morbidity from invasive
bacterial infections during a clinical trial of acellular pertussis
vaccines in Sweden, Pediatr.Infect.Dis.J., 7:637.

Tamura, M., Nogimori, K., Murai, S., Yajima, M., Ito, K., Katada, T., Ui, M.
and Ishii, S., 1982, Subunit structure of the islet-activating protein, pertussis toxin, in conformity with the A-B model, Biochem.,
21:5516.

Uchida, T., Gill, D.M. and Pappenheimer, A.M., Jr., 1971, Mutation in the
structural gene for diphtheria toxin, carried by the temperate phage
beta, Nature New Biol., 233:8.

VACCINATION AGAINST EPSTEIN-BARR VIRUS

M.A. Epstein

Nuffield Department of Clinical Medicine
University of Oxford
John Radcliffe Hospital, Oxford OX3 9DU, UK

INTRODUCTION

Epstein-Barr (EB) virus, one of the six human herpesviruses, was discovered 26 years ago (Epstein et al, 1964) and is of importance in the present context because of its association with certain human cancers. This association was known already in 1976 to be sufficiently close in the case of undifferentiated nasopharyngeal carcinoma (NPC) (Shanmugaratnam, 1971) and endemic Burkitt's lymphoma (BL) (Burkitt, 1963) for it to be suggested that vaccine prevention of EB virus infection might be expected to lead to a decrease in the incidence of the tumours (Epstein, 1976).

RATIONALE FOR A VACCINE AGAINST EB VIRUS

The numerical importance of EB virus-associated NPC in world cancer terms has been discussed in relation to antiviral vaccine control on many occasions, together with the precedent set for such control by the successful intervention against Marek's disease herpesvirus of chickens in order to eliminate Marek's malignant lymphomas in the poultry industry (Churchill et al, 1969; Okazaki et al, 1970). The reasons for selecting the EB virus-determined membrane antigen (MA) and its large glycoprotein component of 340Kd (gp340) as immunogen for use in an EB virus vaccine have also been extensively reviewed pointing out that an analogous antigen from the membranes of cells infected with Marek's herpesvirus is effective as an experimental vaccine against this agent, that EB virus MA has long been known to induce virus-neutralizing antibodies, that MA gp340 does likewise, and that the latest studies have shown that gp340 also elicits T cell responses (Uleato et al, 1988; Bejarano et al, 1990). All these topics have recently been considered again in detail by Morgan and Epstein (1989) and by Epstein (1989).

EARLY WORK TOWARDS A VACCINE BASED ON MA gp340

The preliminary steps were directed towards establishing a colony of the susceptible test animal (the cottontop tamarin, <u>Saguinus oedipus oedipus</u>), devising a sensitive antigen monitoring assay (radio immunoassay), working out with the latter a practical preparative procedure (molecular weight-based), making the product immunogenic (by incorporation in lipo-

somes), and allowing ready quantitation of induced antibodies (by ELISA).
Each of these steps was reported at the time and was subsequently summarized
(Epstein, 1984). Using these methods it proved possible to validate the
ability of a prototype vaccine based on MA gp340 to protect tamarins against
tumour induction by a 100% carcinogenic dose of challenge EB virus (Epstein
et al, 1985).

EB VIRUS VACCINES FOR USE IN MAN

Attempts to make MA gp340 in genetically engineered cells, to use
recombinant vaccinia and varicella viruses engineered to express MA gp340,
and to purify MA gp340 from EB virus carrying cell lines by improved techniques involving fast protein liquid chromatography (FPLC) have all been
pursued and discussed elsewhere (Morgan and Epstein, 1989; Epstein, 1989).
Only the latter has given encouraging results and opened up the possibility
of human trials.

A First Generation MA gp340-based Human Vaccine Preparation

Antigen. During preliminary analysis of the structure of MA gp340 it
was observed that the carbohydrate component was heavily sialylated (Morgan
et al, 1984) and it was concluded that the molecule should therefore remain
negatively charged at relatively low pH. Purification by anion exchange
chromatography on an automated FPLC system was accordingly developed (David
and Morgan, 1988). With this method the membranes of EB virus-infected
cells (expressing MA gp340) were solubilized in a synthetic non-ionic
detergent (MEGA-9) and passed down an ion exchange matrix (MonoQ) under
conditions of low ionic strength at pH5. In these circumstances MA gp340
binds efficiently to the anion exchanger and can be readily eluted in a salt
gradient; although virtual purity is achieved with this single step, a final
gel filtration gives a homogeneous product. The important advantages of
this method are that all MA gp340 molecules are isolated, denaturing is
avoided, automation and reproducibility are inherent, and there is potential
for scale-up. FPLC gp340 provides an excellent and easily made vaccine
immunogen which has been investigated biologically in tamarins when mixed
with a synthetic muramyl dipeptide (MDP) analogue.

Adjuvant. MDP has long been known to have very potent adjuvant activity but it is also pyrogenic and induces such serious side effects as
arthritis, anterior uveitis, and central nervous system pathology. The
immunopotentiating activity has been separated from the functions causing
side effects with the introduction of a synthetic threonyl derivative of MDP
which, when emulsified in squalane with L121 pluronic polymer gives a power-
ful adjuvant formulation (Allison and Byars, 1986 and 1987). On mixing with
immunogens, the formulation consists of microspheres which retain antigen on
their surfaces and activate complement, and both humoral and cellular immunity have been induced with a variety of viruses and viral subunits (Allison
and Byars, 1987). An IND has recently been issued (June 1990) by the US
FDA.

Biological activity. FPLC MA gp340 antigen mixed with threonyl MDP
adjuvant has been assessed for efficacy as a protective vaccine against EB
virus in tamarins. A particular practical advantage of this preparation
lies in the fact that antigen and adjuvant can simply be mixed together
before injection without the need for complicated coupling.

Four colony-bred tamarins received five s.c. injections of vaccine
preparation at two-weekly intervals and blood samples were taken every two
weeks during and after immunization to monitor by ELISA the levels of serum
antibody to gp340; each dose of vaccine preparation contained 50 μg FPLC MA

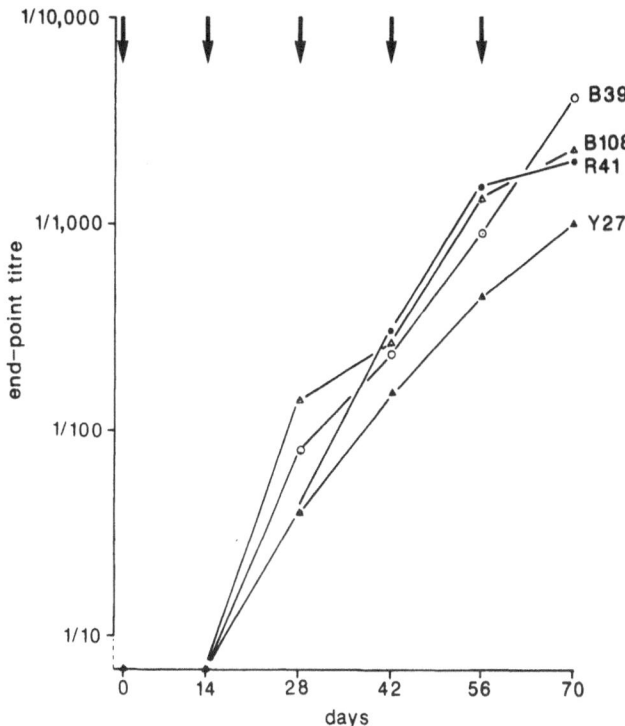

Fig. 1. Antibody responses in four tamarins after immunization with
FPLC-purified MA gp340 in threonyl MDP adjuvant formulation.
ELISA end-point titres are plotted against time in days.
Arrows indicate the points of immunization. (From Morgan et
al, 1989, courtesy of the editor and publishers of the Journal
of Medical Virology.)

gp340. It can be seen from Fig. 1 that the vaccine preparation rapidly and
efficiently induced specific high titre antibody reponses; indeed, even
three inoculations were sufficient for this purpose. In addition, all four
sera taken two weeks after the final dose of vaccine preparation were very
strongly virus-neutralizing.

Two weeks after the last injection the four immunized tamarins and an
untreated control animal were all challenged with a 100% carcinogenic dose
of EB virus ($10^{5 \cdot 3}$ tissue culture transforming units). After fourteen days
the untreated control animal developed multiple, rapidly growing lymphoid
tumours exactly as in all previous experiments, whereas the vaccinated
tamarins remained entirely free of tumours (Fig. 2). Some very minor lymph
node enlargement was observed, but it did not progress, and when an enlarged
node (from tamarin B39) was biopsied and examined histologically, an infil-
trate of reactive proliferating cells was the only abnormality found.
Exactly the same complete protection against the challenge virus has now
been achieved in pilot tamarin experiments using a similar vaccine preparation but with the FPLC MA gp340 content reduced to 5 µg per dose, a significant point in the context of production costs. These results have been
reported in detail (Morgan et al, 1989).

Thus, a vaccine preparation suitable for human use has been shown to be
fully protective against EB virus in experimental animals and the production
methods can be adapted to Good Manufacturing Practice and the requirements
of licensing authorities.

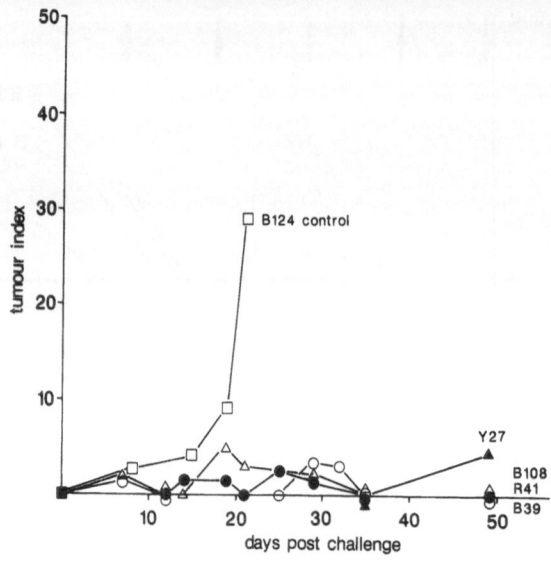

Fig. 2.   Responses of tamarins to a 100% carcinogenic dose of EB virus
(10$^{5.3}$ tissue culture transforming units). The tumour index
(sum of radii of all palpable tumours in each animal) is
plotted against time in days. Rapidly progressing tumours
were detectable in a normal control animal (B124) two weeks
after the challenge, whereas the four vaccinated tamarins
remained free of tumours. (From Morgan et al, 1989, courtesy
of the editor and publishers of the Journal of Medical Virology.)

A PHASE I TRIAL OF FPLC MA gp340

        After appropriate testing for immunogenicity, sterility, and safety
both in vitro and in laboratory animals, a production batch of FPLC MA gp340
and of threonyl MDP adjuvant formulation will be used for a Phase I human
trial. The objectives of such a trial are:-

  i) To determine the immunogenicity in man of FPLC-purified EB virus MA
     gp340 administered with the MDP adjuvant formulation by injection.

 ii) To make a preliminary estimate of its tolerability in particular if
     given to seropositive individuals.

The protocol for the Phase I trial involves the folowing steps:-

  i) Serological screening for antibodies to EB virus of 100 young,
     informed, consenting human volunteers.

 ii) Selection of 12 seronegative individuals and 12 seropositive
     individuals with low levels of antibody to MA gp340.

iii) Full clinical and laboratory screening of those chosen.

 iv) Full investigation of EB virus status of those chosen (serology, saliv-
     ary virus shedding, cytotoxic T cell function, establishment of B cell
     lines as subsequent targets for cell mediated immunity).

  v) Administration of vaccine.

 vi) Post-immunization clinical monitoring and scrutiny of daily diary of
     symptoms.

110

vii) Immunological assessment of antibody and cell mediated responses to EB virus.

The analysis of the results will concentrate on the search for local or systemic reactions to the vaccine, and on the detection in seronegative individuals of induced antibody and specific cellular responses to MA gp340 and in seropositive individuals of a boost in these measures of immunity.

CONCLUDING REMARKS

It is sometimes argued that to attempt to use vaccines in man to prevent infection with cancer-associated viruses in order to decrease the incidence of the tumours cannot be justified in the absence of a clear understanding of the mechanisms by which the viruses cause the cancers, or even certain knowledge that they do so. The successful control of a human cancer by vaccination against its tumour virus, analogous to the antiviral vaccine control of Marek's disease lymphomas (Churchill et al, 1969; Okazaki et al, 1970), is in fact the only way in which causality can be proven in the human context. And as for counsels of delay until mechanisms are understood, they can best be compared to the toleration of heavy smoking until the exact details of the role of cigarettes in the induction of lung cancer have been worked out.

Subunit antiviral vaccines have been highly effective in both human and animal diseases and although it is hoped that the present first generation FPLC MA gp340-based vaccine will soon be superceded by more sophisticated second generation products, it is important that the protective immunogenicity of MA gp340 should be assessed in man in the meantime.

Progress with a vaccine against EB virus continues steadily and success at each step in demonstrating the efficacy of the MA gp340-based vaccine has accelerated the rate of advance. The long-term goals have appeared dauntingly distant for many years but it now seems likely that they will at least be explored sooner rather than later.

REFERENCES

Allison, A.C. and Byars, N.E., 1986, An adjuvant formulation that selectively elicits the formation of antibodies of protective isotype and cell mediated immunity, J.Immunol.Methods,95:157.

Allison, A.C. and Byars, N.E., 1987, Vaccine technology: Adjuvants for increased efficiency, Biotechnol., 5:1041.

Bejarano, M.T., Masucci, M.G., Morgan, A., Morein, B., Klein, E. and Klein, G., 1990,Epstein-Barr virus (EBV) antigens processed and presented by B cells, B blasts, and macrophages trigger T-cell-mediated inhibition of EBV-induced B-cell transformation, J.Virol., 64:1398.

Burkitt,D., 1963, A lymphoma syndrome in tropical Africa, in: "International Review of Experimental Pathology", Vol. 2, G.W. Richter and M.A. Epstein, eds.

Churchill, A.E., Payner, L.N. and Chubb, R.C., 1969, Immunization against Marek's disease using a live attenuated virus, Nature, 221:744.

David, E.M. and Morgan, A.J., 1988, Efficient purification of Epstein-Barr virus membrane antigen gp340 by fast protein liquid chromatography, J.Immunol.Methods, 108:231.

Epstein, M.A., 1976, Epstein-Barr virus - is it time to develop a vaccine program? J.Nat.Cancer Inst., 56:697.

Epstein, M.A., 1984, A prototype vaccine to prevent Epstein-Barr (EB) virus-associated tumours, Proc.Roy.Soc.B.Lond., 221:1.

Epstein, M.A., 1989, Present progress with vaccines against Epstein-Barr virus infection, in: "Immunological Adjuvants and Vaccines", G. Gregoriadis, A.C. Allison, and G. Poste, eds., Plenum Press, New York.

Epstein, M.A., Achong, B.G. and Barr, Y.M., 1964, Virus particles in cultured lymphoblasts from Burkitt's lymphoma, Lancet, i:702.

Epstein, M.A., Morgan, A.J., Finerty, S., Randle, B.J. and Kirkwood, J.K., 1985, Protection of Cottontop tamarins against Epstein-Barr virus-induced malignant lymphoma by a prototype subunit vaccine, Nature, 318:287.

Morgan, A.J. and Epstein, M.A., 1989, Prospects of immunization against EB virus, in: "Recent Developments in Prophylactic Immunization", A.J. Zuckerman, ed., Kluwer Academic Publishers, Dordrecht.

Morgan, A.J., Smith, A.R., Barker, R.N. and Epstein, M.A., 1984, A structural investigation of the Epstein-Barr (EB) virus membrane antigen glycoprotein, gp340, J.Gen.Virol., 65:397.

Morgan, A.J., Allison, A.C., Finerty, S., Scullion, F.T., Byars, N.E. and Epstein, M.A., 1989, Validation of a first generation Epstein-Barr virus vaccine preparation suitable for human use, J.Med.Virol., 29:74.

Okazaki, W., Purchase, H.G. and Burmester, B.R., 1970, Protection against Marek's disease by vaccination with a herpesvirus of turkeys, Avian Dis., 14:413.

Shanmugaratnam, K., 1971, Studies on the etiology of nasopharyngeal carcinoma, in: "International Review of Experimental Pathology", G.W. Richter and M.A. Epstein, eds., Vol. 10, Academic Press Inc., London.

Uleato, D., Wallace, L., Morgan, A., Morein, B. and Rickinson, A., 1988, In vitro T cell responses to a candidate Epstein-Barr virus vacccine: Human CD4+ T cell clones specific for the major envelope glycoprotein gp340, Euro.J.Immunol., 18:1689.

ADENOVIRUS VECTORED VACCINES

K-H. Hsu, M. Lubeck, S. Mizutani, R. Natuk, M. Chengalvala,
P. Chanda, R. Bhat, B. Bhat, B. Mason, B. Selling, B. Kostek,
A. Davis and P. Hung

Wyeth-Ayerst Research
P.O. Box 8299
Philadelphia, PA 19101, USA

INTRODUCTION

Live adenovirus vaccines have been successfully used for prevention of
acute respiratory disease in military recruits in the past two decades
(Couch et al., 1963; Top et al, 1971; Top, 1975). Such vaccines are composed of enteric-coated tablets of adenovirus types 4 (Ad4) and 7 (Ad7).
Asymptomatic infection of the intestine with orally administered vaccines
has been highly effective in inducing immunity against adenovirus respiratory disease. The long history of safe and efficacious use of the vaccines
have prompted us (Davis et al, 1985; Hung et al, 1988; Lubeck et al, 1989;
Hsu et al, 1991) and others (Saito et al, 1985; Ballay et al, 1985, Johnson
et al, 1988; Dewar et al, 1989) to develop a variety of live recombinant
vaccines using adenovirus as vectors.

The adenovirus vectored vaccines currently under development in our
laboratories include hepatitis B vaccine, AIDS vacine and respiratory syncytial virus vaccine.

CONSTRUCTION OF ADENOVIRUS VECTORS

Recombinant adenovirus has a DNA packaging limit. If the recombinant
genome exceeds the normal size by more than about 5%, it will not be
properly encapsidated. Therefore, most recombinant adenovirus expression
vectors contain deletions that allow the recombinant adenovirus genome to
accommodate more foreign gene.

The E3 region of adenovirus genome (Figure 1) has been shown to be non-
essential for growth of the virus in cell culture (Berkner and Sharp, 1983).
Furthermore, deletion of E3 proteins does not markedly affect the course of
infection in an animal model. The E3 region is, therefore, a likely site
for the insertion of the foreign gene, or it may be deleted to permit the
insertion of excess foreign DNA elsewhere in the adenovirus genome. We have
successfully inserted different foreign genes in the E3 region of the viral
genome with or without concomitant deletion of a portion of the E3 region
(Davis et al, 1985; Lubeck et al, 1989) (Fig. 2a). The foreign DNA has been
inserted, either as part of an expression cassette or under the control of

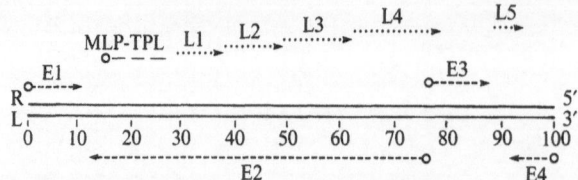

Fig. 1. A simplified diagram of the transcription map of an adenovirus. Each map unit is approximately 360 base pairs.
(o) Represent promoters, (---) indicate early transcription region, (···) indicate late RNA species, MLP, major
late promoter, TPL, tripartite leader. The late RNA species
is derived by alternate RNA splicing from the major late
transcription region. As a result of this, each late mRNA
includes the TPL.

an endogenous adenovirus promoter. An expression cassette system, which has
been frequently used in our laboratories, contains the adenovirus major late
promoter (MLP), the tripartite leader (TPL), followed by the foreign gene
and polyadenylation and processing signals. This expression cassette has
been inserted in, in addition to the E3 region, between the E4 region and
inverted terminal repeat (ITR) of the viral genome (Mason et al, 1990;
Chanda et al, 1989) (Fig. 2b).

ADENOVIRUS-VECTORED HEPATITIS B VACCINES

As a major etiologic agent of infectious liver disease, hepatitis B
virus (HBV) continues to pose significant health problems worldwide.
Although the yeast-derived hepatitis B vaccine has effectively induced
immune responses against HBV infection, this preparation may be too
expensive to allow world wide immunization. Live recombinant hepatitis
vaccines using adenovirus vectors represent an attractive alternative which
offers the advantages of inexpensiveness and easy administration.

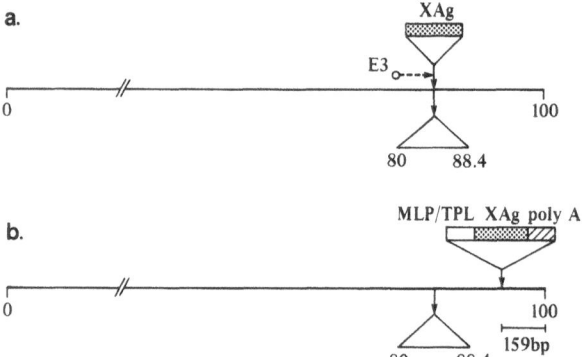

Fig. 2. Structures of two recombinant Ad7 viruses. Both of the
recombinants has a deletion (80-88.4 map unit) in E3 region.
(a) the foreign gene (XAg) is inserted at 80 map unit, which
is behind the E3 promotor. (b) An expression cassette, containing MLP, TPL, XAg and polyA, is inserted in between the
E4 region and right ITR, in these, no transcriptional activity
has been detected.

114

Table 1. Anti-HBsAg Responses in Dogs Following Immunizations with
Ad7-HBsAg/Ad4-HBsAg.

| Group[a] | Anti-HBsAg Titer[b] (mIU) | | | | | |
| | Post-Primary Immunization | | | Post-Booster Immunization[c] | | |
	0d	2w	4w	0d	2w	4w
A	2	150	117	74	21328	10447
B	1	50	8	12	20	122
C	0	2	2	–	–	–

[a]Four dogs per group. Groups A and B were given 1x10^9 PFU and 1x10^8 PFU of
Ab7HBsAg, respectively. Group C was given 1x10^9 PFU of U.V.-irradiated
virus, the residual titer after inactivation was <10^2 PFU.
[b]Data presented are mean titers.
[c]The dogs were booster immunized with Ad4HBsAg at the same dose as primary
immunization, at three months post-primary immunization.

We have made considerable progress in the construction of recombinant
adenoviruses that express large amounts of HBV surface antigen (HBsAg) in
vitro. Some of the early constructed recombinants produce about 1 µg HBsAg
per million cells. In comparison, many of the newly constructed recombinants produce at least a ten-fold greater amount (unpublished data).

Preclinical testing of recombinant adenovirus in animals has been
difficult due to the highly restricted host range of human adenoviruses.
Although the cotton rat (Pacini et al, 1984) and the hamster (Hjorth et al,
1988) support lung replication of various adenovirus serotypes, these
animals are not susceptible to infections by Ad4 or Ad7 (Pacini et al, 1984)
(unpublished data). In our laboratories, a dog model has been established
and routinely used for testing the immunogenicity of Ad4 and Ad7 vectored
vaccines. The data in Table 1 summarize the results of intratracheal inoculation of the dogs with an Ad7-HBsAg followed by booster inoculation with
an Ad4-HBsAg. Anti-HBsAg antibodies were induced by two weeks post-primary
immunization. Following booster immunization, secondary anti-RSV responses
were developed, which reached higher levels than the primary responses
(Chengalvala et al, 1991).

We have also demonstrated that chimpanzees are susceptible to enteric
infection by human Ad7 and Ad4 (Lubeck et al, 1989). Moreover, sequential
oral immunization of two chimpanzees with Ad7-HBsAg induced significant
antibody responses to HBsAg. After challenge with HBV, one chimpanzee was
protected from acute hepatitis and the other chimpanzee experienced modified
HBV-induced disease. These data demonstrate the feasibility of using orally
administered recombinant adenoviruses as a general approach to vaccination.

ADENOVIRUS-VECTORED AIDS VACCINES

AIDS is now a world-wide epidemic for which the development of an
effective vaccine is urgent. Our approach toward the development of an AIDS
vaccine has focused on the expression of the HIV envelope (env) gene using
adenovirus vectors. We have inserted the HIV env gene and the rev gene

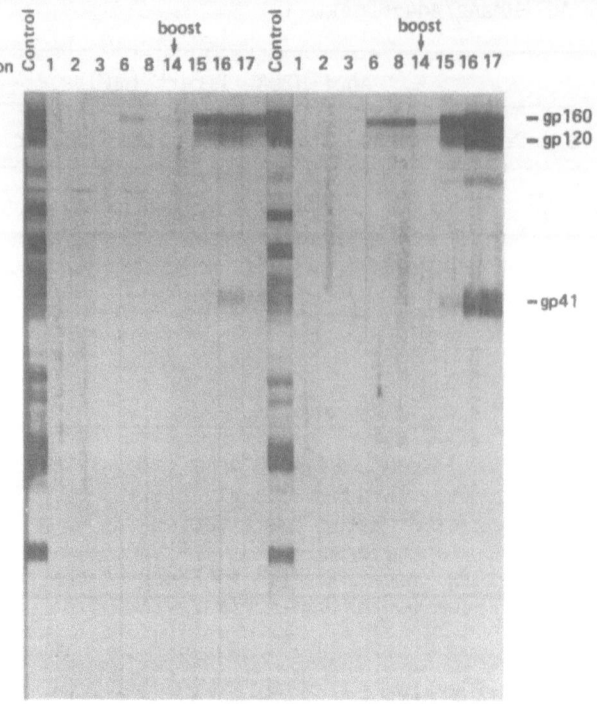

Fig. 3. Western blot analysis of anti-HIV responses in two dogs
following immunizations with Ad7-HIV env rev/Ad4-HIV env rev.
The dogs were inoculated intratracheally with 1x10⁹ PFU of
Ad7-HIV env rev and boosted with 5x10⁹ PFU of Ad4-HIV env
rev, 14 weeks post-primary immunization.

separately in an Ad7 genome. High-level expression of the HIV env gene *in
vitro* has been achieved by co-infection of A549 cells with both Ad7-HIV env
and Ad7-HIV rev (Hung et al, 1988; Chanda et al, 1989).

Accordingly, recombinant adenovirus vectors which contain both the env
and the rev genes were constructed. The viruses expressed high amounts of
env in cell culture (Natuk et al, 1990). Western blot analysis of cell
lysates indicated that the precursor gp160 as well as the cleavage products
gp120 and gp41, were present. The env gene was biologically active as
measured by the ability to induce syncytia formation in HeLa cell cultures
expressing the T4 surface receptor. These recombinants were further tested
in dogs. Intratracheal inoculation of the dogs with an Ad7-HIV env rev
induced anti-gp160 antibodies detectable by Western blot, 6 to 8 weeks post-
primary immunization (Figure 3). The responses were enhanced by booster
immunization with an Ad4-HIV env rev. In addition to anti-gp160, antibodies
to gp120 and gp41 were also detected at 1 to 2 weeks post-booster immunization.

ADENOVIRUS-VECTORED RESPIRATORY SYNCYTIAL VIRUS VACCINE

Human respiratory syncytial virus is one of the leading causes of
severe lower respiratory track disease in infants and young children (Murphy
et al, 1988). Current approaches to vaccine development are centered on

Table 2. Anti-RSV Responses in Dogs Following Immunization with Ad7-RSVF/Ad4-RSVF.

	Anti-RSV Neutralizing Titer[b]					
Group[a]	Post-Primary Immunization			Post Booster Immunization[c]		
	0d	2w	4w	0d	2w	4w
A	<4	<4	<4	<4	32	32
B	<4	23	12	12	113	215
C	<4	10	21	19	682	856
D	<4	<4	<4	<4	<4	<4

[a]Three dogs per group, except group D (2 dogs). Groups A, B and C were given $1x10^6$, $3x10^7$ and $1x10^9$ PFU of Ad7RSVF, respectively. Group D was given $1x10^9$ PFU of U.V.-irradiated virus, the residual titer after irradiation was $<10^2$ PFU.
[b]Data presented are mean titers.
[c]The dogs in Groups A-C were booster immunized with $3x10^8$ PFU of Ad4RSVF, Group D was given the same dose of U.V.-irradiated virus.

subunit vaccines (Brideau et al, 1989; Walsh et al, 1987) and live recombinant viral vectors (Hsu et al, 1991; Olmstead et al, 1988; Collins et al, 1990).

Previous studies by Collins et al (1990) have shown that vaccination of cotton rats with an Ad5 recombinant expressing the fusion glycoprotein of RSV (RSVF) resulted in a significant reduction of RSV replication in lung tissue upon RSV challenge. To extend the study further, we have established the dog as a challenge model and tested the immunogenicity and protective efficacy to Ad7RSVF and Ad4RSVF in dogs (Hsu et al, 1991). Four groups of dogs were immunized with three different doses of Ad7RSVF including a U.V.-irradiated control and boosted with Ad4RSVF seven weeks post-immunization (Table 2). No anti-RSV neutralizing antibody was detected in dogs that received low dose or U.V.-irradiated Ad7RSVF. The other two groups of dogs that received higher doses produced moderate titers of anti-RSV by 2 to 4 weeks post-immunization. Following booster immunization with Ad4RSF, all the dogs developed high titers of anti-RSV neutralizing antibody. These animals were challenged with RSV eleven weeks post-primary immunization and RSV titers in the lungs were determined three days post-challenge. Control dogs had titers 10^3 infectious units/g lung tissue. Vaccinated dogs did not develop detectable levels of virus. The results indicated that the vaccinations effectively protected dogs against RSV infection.

SUMMARY

We have constructed recombinant adenoviruses that express HVsAg, HIV env or RSVF. The recombinant vectors are derived from adenovirus types 4 and 5 which are currently used as live, oral adenovirus vaccines.

The results of animal studies indicated that the inserted genes are expressed in vivo and significant antibdy responses are induced. The results also indicated the advantage of using heterotypic recombinant adenoviruses for booster application.

The chimpanzee studies demonstrated that oral administration of adenovirus vectored vaccines is feasible.

Acknowledgements

We thank M. Hasson for help in the preparation of the manuscript.

REFERENCES

Ballay, A., Levrero, M., Buendia, M.-A., Tiollais, P. and Perricaudet, M.,
    1985, In vitro and in vivo synthesis of the hepatitis B virus surface
    antigen and of the receptor for polymerized human serum albumin from
    recombinant human adenoviruses, EMBO J., 4:3861.

Berkner, K.L. and Sharp, P.A., 1983, Generation of adenovirus by trans-
    fection of plasmids, Nucleic Acids Res., 11:6003.

Brideau, R.J., Walters, R.R., Stier, M.A. and Wathen, M.W., 1989, Protection
    of cotton rats against human respiratory syncytial virus by vaccin-
    ation with a novel chimeric FG glycoprotein, J.Gen.Virol., 70:2637.

Chanda, P.K., Natuk, R.J., Greenberg, L., Mason, B.B., Bhat, B.M., Dheer,
    S.K., Morin, J.E., Molnar-Kimber, K.L., Mizutani, S., Lubeck, M.D.,
    Davis, A.R. and Hung, P.P., 1989, Expression of human immuno-
    deficiency virus envelope glycoproteins by nondefective adenovirus
    vector, in: "Vaccines 89: Modern Approaches to New Vacines Including
    Prevention of AIDS", F. Brown, eds., Cold Spring Harbor Laboratory
    Press, Cold Spring Harbor, N.Y.

Chengalvala, M., Lubeck, M.D., Davis, A.R., Mizutani, S., Molnar-Kimber, K.,
    Morin, J. and Hung, P.P., 1991, Evaluation of adenovirus type 4 and
    type 7 recombinant hepatitis B vaccines in dogs, in: "Vaccine 91",
    in press.

Collins, P.L., Davis, A.R., Lubeck, M.D., Mizutani, S., Hung, P.P., Prince,
    G.A., Camargo, E., Purcell, R.H., Chanock, R.M. and Murphy, B.R.,
    1990, Evaluation of the protective efficacy of recombinant vaccinia
    viruses and adenoviruses that express respiratory syncytial virus
    glycoproteins, in: "Vaccines 90: Modern Approaches to New Vaccines
    Including Presvention of AIDS", F. Brown, ed., Cold Spring Harbor
    Laboratory Press, Cold Spring Harbor, New York.

Couch, R.B., Chanock, R.M., Cate, T.R., Lang, D.J., Knight, V. and Huebner,
    R.J., 1963, Immunization with types 4 and 7 adenovirus by selective
    infection of the intestinal tract, Am.Rev.Resp.Dis., 88:394.

Davis, A.R., Kostek, B., Mason, B.B., Hsiao, C.L., Morin, J., Dheer, S.K.
    and Hung, P.P., 1985, Expression of hepatitis B surface antigen with
    a recombinant adenovirus, Proc.Natl.Acad.Sci.USA, 82:7560.

Dewar, R.L., Natarajan, V., Vasudevachari, M.B. and Salzman, N.P., 1989,
    Synthesis and processing of human immunodeficiency virus type 1
    envelope proteins encoded by a recombinant human adenovirus,
    J.Virol., 63:129.

Hjorth, R.N., Bonde, G.M., Pierzchala, W.A., Vernon, S.K., Wiener, F.P.,
    Levner, M.H., Lubeck, M.D. and Hung, P.P., 1988, A new hamster model
    for adenovirus vaccination, Arch.Virol., 100:279.

Hsu, K.-H., Lubeck, M., Davis, A., Bhat, R., Selling, B., Bhat, B.,
    Mizutani, S. and Hung, P., 1991, Immunogenicity and protective
    efficacy of adenovirus vectored respiratory syncytial virus vaccine,
    in: "Vaccines 91: Modern Approaches to New Vaccines Including
    Prevention of AIDS", F. Brown et al, eds., Cold Spring Harbor
    Laboratory Press, Cold Spring Harbor, N.Y., in press.

Hung, P.P., Morin, J.E., Lubeck, M.D., Barton, J.E., Molnar-Kimber, K.L.,
    Mason, B.B., Dheer, S.K., Jarocki-Witek, V., Kostek, B., Zandle, G.,
    Conley, A.J. and Davis, A.R., 1988, Recombinant adenovirus as a
    vehicle for the HBV surface antigen or HIV envelope protein genes,
    in: "Human Retroviruses, Cancer and AIDS: Approaches to Prevention
    and Therapy", Alan R. Liss, New York.

Johnson, D.C., Goutman, G.C. and Smiley, J.R., 1988, Abundant expression of herpes simplex virus glycoprotein gB using an adenovirus vector, Virology, 164:1.

Lubeck, M.D., Davis, A.R., Chengalvala, M., Natuk, R.J., Morin, J.E., Molnar-Kimber, K., Mason, B., Bhat, B.M., Mizutani, S., Hung, P.P., Purcell, R.H., 1989, Immunogenicity and efficacy testing in chimpanzees of an oral hepatitis B vaccine based on live recombinant adenovirus, Proc.Natl.Acad.Sci.USA, 86:6763.

Mason, B.B., Davis, A.R., Bhat, B.M., Chengalvala, M., Lubeck, M.D., Zandle, G., Kostek, B., Cholodofsky, S., Dheer, S., Molnar-Kimber, K., Mizutani, S. and Hung, P.P., 1990, Adenovirus vaccine vectors expressing hepatitis B surface antigen: Importance of regulatory elements in the adenovirus major late intron, Virology, 177:452.

Natuk, R.J., Chanda, P.K., Dheer, S.K., Mason, B.B., Rogers, P.E., Lubeck, M.D., Mizutani, S., Davis, A.R. and Hung, P.P., 1990, Expression of HIV-1 env gene using a recombinant adenovirus vector, FASEB, 4:A2259.

Pacini, D.L., Dubovi, E.D. and Clyde, Jr., W.A., 1984, A new animal model for human respiratory tract disease due to adenovirus, J.Infect.Dis., 150:92.

Murphy, B.R., Prince, G.A., Collins, P.L., Coelingh, K.V.W., Olmsted, R.A., Spriggs, M.K., Parrott, R.H., Kim, K.-W., Brandt, C.D. and Chanock, R.M., 1988, Current approaches to the development of vaccines effective against parainfluenza and respiratory syncytial viruses, Virus Res., 11:1.

Olmstead, R.A., Buller, R.M.L., Collins, P.L., London, W.T., Beeler, J.A., Prince, G.A., Chanock, R.M. and Murphy, B.R., 1988, Evaluation in non-human primates of the safety, immunogenicity and efficacy of recombinant vaccinia viruses expressing the F or G glycoprotein of respiratory syncytial virus, Vaccine, 6:519.

Saito, I., Oya, Y., Yamamoto, K., Yuasa, T. and Shimogo, H., 1985, Construction of nondefective adenovirus type 5 bearing a 2.8-kilobase hepatitis B virus DNA near the right end of its genome, J.Virol., 54:711.

Top, Jr., F.H., 1975, Control of adenovirus acute respiratory disease in U.S. Army trainees, Yale J.Biol.Med., 48:185.

Top, Jr., F.H., Buescher, E.L., Bancroft, W.H. and Russell, P.K., 1971, Immunization with live types 7 and 4 adenovirus vaccines, II. Antibody response and protective effect against acute respiratory disease due to adenovirus type 7, J.Infect.Dis., 124:155.

Walsh, E.E., Hall, C.B., Briselli, M., Brandriss, M.W. and Schlesinger, J.J., 1987, Immunization with glycoprotein subunits of respiratory syncytial virus to protect cotton rats against viral infection, J.Infect.Dis., 155:1198.

VACCINES AGAINST BACTERIAL INFECTIONS OF CHILDREN

P. Helena Makela, Helena Kayhty, Aino K. Takala,
Heikki Peltola and Juhani Eskola

National Public Health Institute
SF 00300 Helsinki
Finland

INTRODUCTION

There are four old and well established vaccines for bacterial infect-ions in children: the vaccines against diphtheria, tetanus, tuberculosis and whooping cough (pertussis). Their value has been amply proven and they are now the mainstay of the Expanded Programme of Immunization (EPI) of the WHO, recommended to be given to each infant born in the world. Their actual usage covers roughly two thirds of the world's children and is rapidly increasing through collaborative efforts of EPI. The EPI framework would give an excellent opportunity to introduce new vaccines to rapidly benefit children in all countries. Do we have such vaccines ready? For which diseases would new vaccines be needed (Institute of Medicine, 1986; Griffiss et al, 1987; Robbins and Freeman, 1988)?

A new group of bacterial vaccines, based on the capsular polysacchar-ides of encapsulated bacteria, were introduced in the 1970s (one of them, the pneumococcal vaccine, had been marketed in the 1940s, but was over-shadowed by the advent of penicillin). These polysaccharide vaccines were aimed at preventing meningococcal meningitis or pneumococcal pneumonia. Their efficacy in children was, however, not satisfactory. Thus they could not be considered within the EPI concept and their overall use has been limited. Continuing efforts at improving the immunogenicity of polysacchar-ides especially in infancy have now borne fruit; the first polysaccharide-protein conjugate vaccine has been proven protective in young infants and the way seems open for a set of other important vaccines utilizing the same principle. Most of this article will be concerned with these vaccines which, if widely administered within the EPI programme, could prevent most of the invasive bacterial infections and also a large part of serious res-piratory infections, one of the two main causes of death of children in developing countries.

The other, equally important cause of death of children in the same areas, is diarrhoea. Like respiratory infections, it can be caused by many microbial agents, of which viruses (especially rotavirus) and bacteria like Vibrio cholerae, toxigenic Escherichia coli, Shigella, Salmonella and Campylobacter are the most important. Much effort has been invested in developing vaccines against these agents, but the success is still not cer-tain. The nature of the mucosal infection, possibly requiring local anti-

Vaccines, Edited by G. Gregoriadis *et al.*
Plenum Press, New York, 1991

body production for prevention, is one problem. Oral administration of potential vaccines is therefore considered important and has shown promise in a large field trial with a cholera vaccine consisting of a mixture of whole killed bacteria and the B subunit of cholera toxin (Clemens et al, 1988a, b). The cholera toxin component in the vaccine apparently also gives protection against toxigenic E. coli. Equally promising results were initially obtained with an oral live (attenuated) Salmonella Typhi vaccine (Wahdan et al, 1982), but subsequent studies in areas with a high disease prevalence have been less encouraging (Levine et al, 1987a; Ferreccio et al, 1989). New attenuated salmonella vaccines are being tested (Levine et al, 1987b), and an attenuated Shigella vaccine is on the way (Lindberg et al, 1990). The capsular polysaccharide of Salmonella Typhi has also given approximately 70% protection from typhoid fever (Acharya et al, 1987). A common feature of these vaccines is that their effect is less good and of shorter duration in children than adults and most have not yet been tested in infants. A protective efficacy better than the 60-70% found in several field trials would also be highly desirable. Thus it is not yet possible to seriously discuss these vaccines in the EPI, within which infants must be immunized at a very early age.

VACCINES AGAINST HAEMOPHILUS INFLUENZAE TYPE B

The Need of the Vaccine: Haemophilus influenzae Type b Disease

Haemophilus influenzae type b (Hib) is the most frequent cause of invasive bacterial infections of children less than five years old in the industrialized world, where the annual incidence of Hib meningitis varies from 40 to 60/100,000 (for review, see Ward and Cochi, 1988). Much higher incidence rates are seen in several native populations, e.g. the Apache Indians or Alaskan Eskimos (Losonsky et al, 1984; Ward et al, 1981). The most common manifestation of the disease is meningitis; others include epiglottitis, purulent arthritis, cellulitis, pneumonia and bacteremia; pneumonia may be more common in developing countries (Ward et al, 1981; Takala et al, 1989; Todd and Bruhn, 1975; Shann et al, 1984; Takala, 1989).

From the point of view of vaccination, the age at which the disease occurs is of utmost importance. Although in all areas the disease primarily affects children under five years of age, there are large variations in the proportion of cases occurring under two years of age: approximately 60% of Hib meningitis in Scandinavian countries compared to over 90% among the Indians or Eskimos (Takala, 1989). This limit is important since the immune response to the polysaccharide vaccine (see later) only develops around two years of age. Another important limit is about six months, before which time it is difficult to build up protection with most vaccines. Again, Scandinavian countries fare well in this regard with almost no cases of Hib meningitis in less than six month old infants, whereas over 30% of cases among Eskimos occur before six months of age.

Hib infection is acquired by droplet infection from patient carriers of Hib. Four to five percent of children and adults carry Hib bacteria in their pharynx (Michaels et al, 1976). The carriage rate is much increased in close contacts of a patient, eg. family members or children in the same day care center (Band et al, 1984; Granoff and Basden, 1980). Thus large family size and care in large groups are important risk factors for Hib disease (Takala, 1989).

Hib Polysaccharide Vaccines

Hib bacteria have a polysaccharide capsule that protects them from phagocytosis and lysis by complement attack. Almost all H. influenzae strains isolated from invasive infections are encapsulated and have the same type b capsule, a polymer of ribose-ribitol-phosphate, often abreviated as PRP (Crisel et al, 1975). Thus an obvious approach at preventing Hib infections was a vaccine made of the isolated capsule. This approach was further encouraged by the fact that susceptibility to Hib disease in a population was shown to be inversely related to the presence of bacteriocidal anti-Hib (Fothergill and Wright, 1933) or anti-PRP (Peltola et al, 1977a) antibodies. Normally these antibodies are received transplacentally from the mother, decline in the first months of life and remain at a low level until two years of age. After that they increase to the adult level of about 0.15 ug/ml by 5-6 years of age (Peltola et al, 1977a).

The PRP vaccine was initially shown to be safe, with very few adverse reactions and immunogenic in adults and two year old children (Peltola et al, 1977a; Robbins et al, 1973). The antibody response was, however, much lower in infants and good responses to concentrations seen in unvaccinated adults (who are almost fully immune to the disease) were not seen before two years of age (Peltola et al, 1977a; Peltola et al, 1984).

Protection from invasive Hib disease (Hib isolated from blood and/or cerebrospinal fluid) was monitored in a randomized double blind trial involving approximately 110,000 children between the ages of three months and five years, of whom half received the PRP vaccine and the other half an unrelated meningococcal group A vaccine (Peltola et al, 1977a Peltola et al, 1984). Results showed that the vaccine was 90% protective in children older than 18 months, but failed to protect those younger than this. The total number of patients was too small to determine the age limit of protective action with accuracy, but the overall correlation of protection with immunogenicity was apparent. When the vaccine was then licenced for use in the USA, the recommended age of vaccination was set at two years. The subsequent experience based on case-control studies showed lower figures for efficacy: 44-88% in three studies, 58% in one (reviewed in Ward et al, 1988; Daum et al, 1988). The reasons for this are not clear, but certainly the case-control method is less accurate, especially when risk factors of the disease are not well known and therefore cannot be controlled. It is also possible that the age at which the maturation of the immune response with regard to the polysaccharide vaccine takes place, varies to some extent even in populations in industrialized countries (Anderson and Insel, 1988).

However, even without these uncertainties about the use of PRP it was clear that an Hib vaccine that could not protect children during the most susceptible age period, between six and eighteen months, was unsatisfactory. As a result, other countries did not adopt the PRP vaccine but preferred to wait for an improved vaccine that could be used in infancy.

The lessons learned from the use of the Hib polysaccharide vaccine have certainly been helpful in guiding further work on vaccine development. They helped to identify the main cause of the low immunogenicity of the vaccine as slow maturation of the immune system in respect of polysaccharide antigens. This was attributed to the T cell independent (TI) nature of the antigens and directed further work towards converting the TI polysaccharide to a T cell dependent (TD) form. Interim studies were performed to assess the possibility of augmenting the immunogenicity of PRP by the adjuvant action of pertussis vaccine, but the initial promise (King et al, 1981) did not carry through in subsequent work (Kayhty et al, 1987). Furthermore, selecting for higher molecular weight of the PRP, or its noncovalent complexation with a protein did not look promising even in animal experiments (Schneerson et al, 1980). Covalent conjugation to protein carriers (Schneerson et al, 1980; Chu et al, 1983) did, however, produce the hoped for result.

Hib Conjugate Vaccines

When PRP was conjugated to proteins (several were initially tested in animal experiments) it could be shown to have gained TD properties missing from PRP (Schneerson et al, 1980; Chu et al, 1983). These included a higher proportion of IgG class in the response and priming for an increased secondary response. On the basis of these encouraging findings, the first conjugates were tested for safety and immunogenicity in adults and two year old children and after promising results in these age groups, also in infants (Anderson et al, 1986). The key finding was an antibody response in infancy after three doses of the conjugate that was clearly higher than the response to PRP at the same age. The fact that the response was seen only after three doses (or, with somewhat different dosage schedules, after two doses given two months apart) supported the notion that the conjugate behaved like a TD antigen. Subsequent studies have provided further support to this (see later).

The first conjugate that became ready for a field trial in infants was PRP-D, in which PRP is conjugated via a short linker of adipic acid dihydrazide to diphtheria toxoid (Gordon, 1984). Two such trials were initiated at about the same time, one in Finland where the polysaccharide vaccine had been tested ten years previously, and one among Alaskan Eskimos (Eskola et al, 1985; Eskola et al, 1987; Ward et al, 1990). In the Finnish trial, approximately 55,000 infants received the conjugate at three, four and six months of age, randomized by day of birth. Before the booster dose given at 14-18 months together with the measles, mumps and rubella vaccine (MMR), four cases of invasive Hib disease occurred among the vaccinated children and 37 in the control group, indicating 90% protection (Eskola et al, 1985). After booster, no cases have been seen in the vaccinated children and 27 among the controls. The vaccine thus worked extremely well; this was the first time that it has been possible to prevent Hib infections in infancy. In the Alaskan trial, there was no significant protection (Ward et al, 1990). However, the number of infants in the study was much smaller, they were vaccinated at an earlier age and they came from a population with a high incidence of Hib disease in very young infants - that is a population in which prevention was expected to be especially difficult to achieve.

Other Hib conjugates have utilized different carrier proteins (a mutant non-toxic diphtheria toxin called CRM 197, tetanus toxoid or meningococcal outer membrane protein vesicles), several alternative methods for conjugation and the polysaccharide component of somewhat different molecular weights (Anderson et al, 1985; Claesson et al, 1988; Einhorn et al, 1986). They all differ to some extent in their immunogenic properties (Anderson et al, 1985; Claesson et al, 1988; Einhorn et al, 1986; Kayhty et al, 1989). PRP-D, the first conjugate vaccine, gives the lowest antibody concentrations measured after the basic immunization schedule (two or three doses) in infancy. The conjugates with tetanus toxoid or meningococcal outer membrane protein vesicles differ from the others by inducing relatively high concentrations of anti-PRP already after the first dose.

Although protective efficacy of the conjugates does not seem to be as clearly correlated with serum antibodies (Eskola et al, 1985) as was protection by the polysaccharide vaccine (Peltola et al, 1984; Anderson, 1984; Kayhty et al, 1983), it is reasonable to expect that a higher concentration of antibodies in general may be associated with a higher level of protection. This may become relevant in areas with a high infection pressure such as Alaska: maybe another conjugate would have performed better in that trial than PRP-D, although the latter was quite satisfactory in Finland where the disease incidence is lower. Furthermore, the vaccines that induce an antibody response already with the first dose (therefore at an earlier age than achievable by vaccines requiring two doses sufficiently far apart) may have an advantage in areas where the disease occurs in young infants.

THE T CELL-DEPENDENT NATURE OF CONJUGATE VACCINES

The main advantage sought when developing the Hib conjugate vaccines was immunization and protective ability early in infancy. This was clearly achieved: immunization can be started as early as two months of age and satisfactory antibody concentrations (as high as in unimmunized adults) achieved at an age of three to five months. This is per se, however, not evidence for T cell dependence since the reason for the poor immunogenicity of polysaccharides in infancy is not known and was only suspected to be due to lack of capacity to respond to TI antigens.

Antibody Isotypes in the Response

The immunoglobulin isotype distribution is known in animals to differ between TI and TD responses. By analogy, one would expect in humans a greater percentage of the antibodies to a TD antigen to be of the IgG class and more specifically of IgG1 and IgG3 isotypes. This is the case in the adult booster response to tetanus toxoid or primary response to live rubella vaccine (Seppala et al, 1984; Sarnesto et al, 1985). It is not, however, as simple to compare responses to Hib conjugate vaccine to these. The Hib vaccination in adults would not represent a clean primary or secondary response, because all adults have pre-existing anti-PRP antibodies, presumably induced by mucosal carriage of Hib or other, crossreacting bacteria; the immunizing antigen in them would have been a TI rather than TD form. In infants, on the other hand, the anti-PRP concentration after the primary response is too low for an accurate assessment of its isotype constitution. This problem does not apply to the secondary response at 14-18 months, but still another problem remains; IgG2 isotype appears rather late in the ontogenesis of human infants, around two years of age and thus its absence from the response cannot be taken as an argument for a TD response.

With all these uncertainties, the conclusions that can be drawn from isotype determination of anti-PRP after immunization with Hib conjugate vaccines are that the response is closer to a TD type response than the response to PRP at the same age, but less strictly TD than responses to protein antigens. Thus the conjugate seems to be somewhere in between a TI and a TD antigen when assessed by the isotype distribution of the response, be it the proportion of IgG or the relative proportion of IgG_2 in the anti-PRP antibodies (Makela et al, 1987; Kayhty et al, 1988; Seppala et al, 1988). Furthermore, comparison of two conjugates did not show conspicuous differences between them; however, there was a suggestion that a higher antibody response was associated with a more TD type response (Seppala et al, 1988).

Immunologic Memory

The ability to utilize T cell help and thus generate immunologic memory in both T and B cells is a basic characteristic of a TD antigen. The conjugate vaccines, as well as other experimental polysaccharide-protein conjugates, pass this test in mice and other experimental animals (Schneerson et al, 1980; Chu et al, 1983). It was, however, very important to show that this is also true in humans. It is. Children who received PRP-D at three, four and six (or at four and six) months of age had at seven months of age much higher serum anti-PRP concentrations than did children who received their first dose of PRP-D at six months or who had received several doses of PRP (Kayhty et al, 1984; Kayhty et al, 1987). In fact, several doses of PRP at two to six month intervals did not cause any increase of anti-PRP and in some cases resulted in a slight decrease (Peltola et al, 1977a; Kayhty et al, 1984). The secondary response to PRP-D was even clearer when the children, who had their primary immunization with PRP-D at three, four and six months of age, were given a booster injection at 14-18 months. Then, the antibody concentration attained was higher than ever seen after PRP or a first dose of PRP-D at this age (Kayhty et al, 1989). Even those children who had apparently not responded to the primary immunization (undetectable anti-PRP at seven months of age) now showed a higher booster-type response (Eskola et al, 1985).

The ability to give rise to memory and thus to a typical secondary response with a high antibody concentration probably explains why PRP-D gave such a good protection in the Finnish efficacy trial although the actual antibody concentrations in a fairly large part of the seven month old infants after the full course of three injections were low. Thus the estimations of a serum antibody concentration required to predict protection after a polysaccharide vaccine (Anderson, 1984; Kayhty et al, 1983) would not apply to the immunity induced by a conjugate vaccine. The practical question then arises: how should one assess the protective ability of a conjugate vaccine, short of performing a large field trial for efficacy? On the basis of the experience accumulated from the Finnish trials and reports from other studies, we suggest that the response to a booster dose of vaccine would give a good prediction of protection. The booster dose would, in fact, serve as a challenge to the immune system, in many ways similar to the challenge posed by actual infection and the ability to respond rapidly and efficiently to the former would predict ability to respond to the latter as well. In fact, children immunized with PRP-D have been tested by a booster dose of PRP, which is very close to the PRP capsule on Hib bacteria and the response has been almost the same as that to a booster of PRP-D (data to be published).

Effect of Vaccine on Mucosal Carriage

The Hib polysaccharide vaccine did not have any effect on the naso-pharyngeal carriage rate of Hib bacteria (Peltola et al, 1977a). This was not surprising, since a similar lack of effect, or a marginal effect only, had been observed with other polysaccharide vaccines (Gotschlich et al, 1969a; Greenwood et al, 1978; Herva et al, 1980; Rosen et al, 1984; Koskela, 1986; Pichichero and Insel, 1983). It was concluded that antibodies to the capsular polysaccharides would not prevent colonization or that their concentration on the mucosal surfaces would be too low for such prevention, possibly because the parental immunization might not induce a local antibody production. Nevertheless, Koskela et al (1986) had shown that parenterally administered pneumococcal polysaccharide vaccine did induce antibodies also in the middle ear and Pichichero and colleagues showed a local antibody production in the oropharynx in response to the same vaccine (Pichichero and Insel, 1983).

The ability of a vaccine to reduce mucosal carriage of the pathogen in question is very important for the development of herd immunity, i.e. immunity in the population that also protects unimmunized persons. We have therefore tested for the possible effect of the Hib conjugate vaccines on oropharyngeal carriage of Hib bacteria (Takala et al, 1990). While the carriage rate of Hib was about 4% in unimmunized three year old children, it was reduced to zero in children of the same age who had received an Hib vaccine; at the same time the carrier rates for pneumococci and non-capsulated H. influenzae remained unaltered. We believe that the better effect of the conjugate than that of the polysaccharide Hib vaccine on mucosal carriage is associated with the higher serum (and consequently also mucosal) antibody concentrations induced by the former and/or the immunologic memory that allows a rapid antibody response upon contact with the bacteria and subsequent prevention of colonization.

The apparent ability of an Hib conjugate vaccine to eliminate pharyngeal carriage of the bacteria would be important to verify in other trials. It could mean an additional advantage of conjugate vaccines that would potentiate their protective efficacy. This would be even more important for vaccines against non-invasive infections in which local events on the mucosa of either the respiratory tract or the intestine play a major role.

FUTURE CONJUGATE VACCINES

Vaccines against Neisseria Meningitidis Groups A and C

Polysaccharide vaccines to the group A and C meningococci were developed in the early 1970s mainly for the benefit of Armed Forces (Artenstein et al, 1970; Gotschlich et al, 1969b). They were soon shown efficacious in adults. However, the group C vaccine was much like PRP, a poor immunogen in young children (Gold et al, 1975). The group A vaccine was more promising also in infants: in Finland it appeared to be protective as early as at three months of age, although the number of cases in the youngest infants was small even among the controls (Peltola et al, 1977b). Nevertheless, both immunogenicity and protection data indicated that the group A meningococcal polysaccharide vaccine was different from the group C or PRP vaccines by being active already in early infancy.

On the basis of the above it seemed that the group A meningococcal vaccine could be given within the EPI programme to prevent the recurrent severe epidemics of meningitis caused by group A meningococci especially in the "meningitis belt" in Africa (Lapeyssonnie, 1963; Greenwood and Wali, 1980). This proposal was not accepted and the vaccine is only used to combat epidemics by trying to intervene as early as possible when an epidemic starts - a process that would require careful epidemiological surveillance and effective implementation of nationwide vaccination programmes, both very difficult in any developing country.

Follow-up studies of such use of the group A vaccine have, however, shown a major shortcoming in the vaccine in respect of its use in the EPI programme, namely the relatively short duration of its protectice ability (Reingold et al, 1985). This is somewhat in contrast to the experience in Finland, where protection was shown to last at least five years and the elevated serum antibody concentrations to be even longer (Peltola, 1987; Kayhty et al, 1980). The difference may, again, reflect differences in the infection pressure: in Finland with a low disease incidence, relatively low antibody concentrations would suffice to prevent infection, whereas in Upper Volta higher antibody concentrations would be required.

The relatively short duration of the effect of the polysaccharide vaccines at best (as exemplified by the group A meningococcal vaccine), combined with a general lack of immunologic memory and poor responsiveness of infants (to the group C meningococcal vaccine) all speak for a need for improved vaccines. Conjugate vaccines derived from these polysaccharides would be a logical solution (Griffiss et al, 1987). Technically they should be easy to prepare and early animal experiments have been promising (Beuvery et al, 1982; Jennings and Lugowski, 1981). What is needed is a speedy assessment of the antibody concentrations and booster responses (as discussed for the Hib conjugates above) after administration of these vaccines within the context of the EPI programme. How soon protective efficacy can be ascertained would depend on the epidemic situation in the trial site. The previously demonstrated correlation of serum antibodies with protection, combined with the knowledge of protective efficacy of the Hib conjugate vaccines even in the presence of low antibody concentrations but associated with a good booster response, should lead to a rapid adoption of the meningococcal conjugates in EPI even before protection of disease has been shown. The expected herd immunity mediated by reduced carriage associated with the use of the conjugate vaccines would be very important in respect of meningococcal epidemics.

Pneumococcal Vaccines

Streptococcus pneumoniae, the pneumococcus, may be the greatest killer of people in all times, areas and age groups. It is the most common single agent causing respiratory infections like pneumonia, sinusitis and otitis media; it is also an important cause of bacteremic infections and meningitis in infants and children and also in elderly people (Klein, 1981; Shann, 1986). Respiratory infections on the other hand are the most common infections at all ages and in all areas of the world; in developing countries they are, together with diarrhoea, the most common cause of death in infants and children, whereas in industrialized countries they are perhaps equally common but less deadly. Nevertheless, the pneumonia that leads to the death of old people is most often pneumococcal and this is also true of the excess pneumonia and death during influenza epidemics. A vaccine to prevent pneumococcal infection would therefore be of great importance, even more so with the emergence and increase of antibiotic resistance among the pneumococci.

All pneumococci causing infections are encapsulated. Thus, a vaccine based on the capsular polysaccharide would be a logical way of prevention. A problem is posed by the many varieties (serotypes) among the capsular polysaccharides, but this can be overcome by incorporating all the common serotype polysaccharides in the vaccine. This has been done on the basis of extensive epidemiologic studies and the present polyvalent pneumococcal vaccine covers 23 types and about 90% of pneumococcal infections (Robbins et al, 1983).

The capsular polysaccharide vaccine has been shown to protect adults from pneumonia as well as from bacteremic infections caused by those pneumococcal types represented by the vaccine (Austrian et al, 1976; Bolan et al, 1986). Many of the serotype polysaccharides are, however, poorly immunogenic in infancy (Leinonen et al, 1986) and thus the vaccine had only a marginal effect on otitis media (Karma et al, 1985) or on respiratory infections of children in Papua, New Guinea (Riley et al, 1981).

A conjugate vaccine in which the most common pneumococcal polysaccharides, or at least those common in children, were conjugated to a protein carrier, would be expected to solve this problem. The technology for conjugation exists and has been shown to apply to a pneumococcal polysaccharide (Chu et al, 1983). The experience with the Hib conjugates gives a firm basis for expecting that such conjugates would be immunogenic in infants. The ability of the Hib conjugate vaccine to eliminate pharyngeal carriage is an additional optimistic signal for prevention of even mild and local pneumococcal infections on mucosal sites in the ear, in the paranasal sinuses and in the lungs. The concomitant herd immunity might be especially important for the common forms of respiratory infections.

Vaccine Against Group B Streptococci

Streptococci of group B are an important cause of neonatal septic infections, in which the infection is usually acquired from the birth canal. These bacteria are also encapsulated and antibodies to the capsule (acquired transplacentally) are protective. It would therefore seem logical to immunize the mothers to prevent these serious infections. Again, a capsular

poly-saccharide vaccine has been prepared and tested for immunogenicity
(Baker et al, 1988; Baker, 1990). The results of the immunogenicity study
were not as good as hoped for: a fairly large proportion of the mothers did
not respond. The low response rate has been suggested to be due to genetic
inability to respond to this polysaccharide. If so, this problem would also
be expected to be solved by a conjugate vaccine. Quite a different problem,
but also a serious one, is how to overcome the cultural prejudice towards
immunization during pregnancy. However, it should be remembered that tetanus toxoid given to pregnant mothers is part of the EPI programme in all
developing countries.

Vi Vaccine for Typhoid Fever

As mentioned in the Introduction, the capsular Vi polysaccharide of
Salmonella Typhi has been recently introduced as a promising vaccine candidate to prevent typhoid fever (Acharya et al, 1987). To make its use
possible within EPI, this polysaccharide should also be converted to a
protein conjugate that could immunize infants and induce a long-lasting
memory. First experience with such conjugates is, in fact, promising (Szu
et al, 1987; Szu et al, 1989).

POSSIBLE PROBLEMS WITH CONJUGATE VACCINES

The above gives an idea of the expected expansion with conjugate vaccine technology. Further away and of less clear cut general importance are
possible applications to the capsular polysaccharides of many E. coli and
Klebsiella serotypes causing urinary tract infections and bacteremic infections especially in immunocompromised persons, or the so called O-antigenic
polysacchaaride part of lipopolysaccharide of many enteric bacteria or of
Pseudomonas aeruginosa. Of general use, but of uncertain success at the
present state of knowledge, would be conjugates of the short-chain lipopolysaccharide of meningococci or of Haemophilus influenzae.

A special problem arises with the capsular polysaccharides that are
structurally related to glycoconjugates of the human body (Jann and Jann,
1985). These polysaccharides are very weak immunogens and those antibodies
that are formed are of IgM class and of low affinity (Leinonen and Frasch,
1982). The best known example may be the capsule of group B meningococci
(identical to the K1 capsule of E. coli) that has structural and serological
similarity with glycopeptides in the developing brain and also regenerating
nerve and muscle tissues (Finne et al, 1983; Saukkonen et al, 1986).
Because of such crossreactivity it may be dangerous to try to overcome this
immunological barrier by conjugating the polysaccharide to protein.

An easier problem with the expected expansion of conjugate vaccines is
the question of the carrier protein. How many conjugates could/should use
the same carrier? The first conjugates were made using a well-known and
generally accepted vaccine protein as carrier - this was useful to facilitate their use in the experimental stage and acceptance for childhood
immunization programmes. It is, however, thought that if many conjugates
used the same carrier this would result in too high concentrations of antibodies to the carrier with ensuing adverse effects. So far there are, however, no observations suggesting that this would have occurred - in fact all
conjugate vaccines have been very well tolerated.

The possible need of new carrier proteins has been anticipated in one
Hib conjugate that uses meningococcal outer membrane protein vesicles as the
carrier (Einhorn et al, 1986). Cholera toxoid has also been used as carrier
(Szu et al, 1989) and might serve as a basis for an orally administered
conjugate vaccine. However, a more general approach in the long run might

be to conjugate the polysaccharide of a certain bacterial species to a
purified protein from the same species. An additional advantage might be
gained if the protein itself could also induce protection. An interesting
possibility would be, for example, a conjugate of pneumococcal polysaccharides to a nontoxic form of pneumolysin, a protein toxin shared by all pneumococci that is at least somewhat protective in animal experiments (Paton et
al, 1983).

REFERENCES

Acharya, V.L., Lowe, C.U., Thapa, R., Gurubacharya, V.L., Shrestha, M.B.,
    Cadoz, M., Schulz, D., Armand, J., Bryla, D., Trollfors, B., Gramton,
    T., Schneerson, R. and Robbins, J.B., 1987, Prevention of typhoid
    fever in Nepal with the Vi capsular polysaccharide of Salmonella
    typhi. A preliminary report, N.Engl.J.Med., 317:1101.
Anderson, P., 1984, The protective level of serum antibodies to the capsular
    polysaccharide of Haemophilus influenzae type B, J.Infect.Dis.,
    149:1034.
Anderson, P. and Insel, R.A., 1988, Prospects for overcoming maturational
    and genetic barriers to the human antibody response to the capsular
    polysaccharide of Haemophilus influenzae type b, Vaccine, 6:188.
Anderson, P.W., Pichichero, M.E., Insel, R.A., Betts, R., Eby, R. and Smith,
    D.H., 1986, Vaccines consisting of periodate-cleaved oligosaccharides
    from the capsule of Haemophilus influenzae type b coupled to a pro-
    tein carrier: structural and temporal requirements for priming in the
    human infant, J.Immunol., 137:1181.
Anderson, P., Pichichero, M.E. and Insel, R.A., 1985, Immunogens consisting
    of oligosaccharides from the capsule of Haemophilus influenzae type b
    coupled to diptheria toxoid or CRM197, J.Clin.Invest., 76:52.
Artenstein, M.S., Gold, R., Zimmerly, J.G., Wyle, F.A., Schneider, H. and
    Harkins, C., 1970, Prevention of meningococcal disease by group C
    polysaccharide vaccine, N.Engl.J.Med., 282:417.
Austrian, R., Douglas, R.M., Schiffman, G., Coetzee, A.M., Koornhof, H.J.,
    Hayden-Smith, S. and Reid, R.D.W., 1976, Prevention of pneumococcal
    pneumonia by vaccination, Trans.Assoc.Am.Physicians, 89:184.
Baker, C.J., 1990, Immunization to prevent group B Streptococcal Disease:
    victories and vexations, J.Infect.Dis., 161:917.
Baker, C.J., Rench, M.A., Erwards, M.S., Carpenter, R.J., Hays, B.M. and
    Kasper, D.L., 1988, Immunization of pregnant women with a poly-
    saccharide vaccine of group B streptococcus, N.Engl.J.Med.,
    319:1180.
Band, J.D., Fraser, D.W. and Ajello, G., 1984, Haemophilus influenzae
    disease study group. Prevention of Haemophilus influenzae type b
    disease, JAMA, 251:2381.
Beuvery, E.C., van Rossum, F. and Nagle, J., 1982, Comparison of the
    induction of immunoglobulin M and G antibodies in mice with purified
    pneumococcal type 3 and meningococcal group C polysaccharides and
    their protein conjugates, Infect.Immun., 37:15.
Bolan, G., Broome, C.V., Facklam, R.R., Plikaytis, B.D., Fraser, D.W. and
    Schlech, W.F., 1986, Pneumococcal vaccine efficacy in selected
    populations in the United States, Ann.Intern.Med., 104:1.
Chu, C.Y., Schneerson, R., Robbins, J.B. and Rastogi, S.C., 1983, Further
    studies on the immunogenicity of Haemophilus influenzae type b and
    pneumococcal type 6A polysaccharide-protein conjugates,
    Infect.Immunol., 40:245.
Claesson, B.A., Trollfors, B., Lagergard, T., Taranger, J., Bryla, D.,
    Otterman, G., Cramton, T., Yang, Y., Reimer, C.R., Robbins, J.B. and
    Schneerson, R., 1988, Clinical and immunological responses to the
    capsular polysaccharide of Haemophilus influenzae type alone or
    conjugated to tetanus toxoid in 18 to 23 month old children,
    J.Pediatr., 112:695..

Clemens, J.D., Sack, D.A., Harris, J.R., Chakraborty, J., Khan, M.R., Stanton, B.F., Ali, M., Ahmed, F., Yunus, M. and Kay, B.A., 1988a, Impact of B subunit killed whole-cell and killed whole-cell-only oral vaccines against cholera upon treated diarrheal illness and mortality in an area endemic for cholera, Lancet, i:1375.

Clemens, J.D., Sack, D.A., Harris, J.R., Chakraborty, J., Neogy, P.K., Stanton, B., Huda, N., Khan, M.U., Kay, B.A., Khan, R., Ansaruzzaman, M., Yunus, M., Rao, M.R., Svennerholm, A-M. and Holmgren, J., 1988b, Cross-protection by B subunit whole-cell cholera vaccine against diarrhoea associated with heat-labile toxin producing enterotoxigenic Escherichia coli: results of a large scale field trial, J.Infect.Dis., 158:372.

Daum, R.S., Marcuse, E.K., Giebink, G.S., Hall, C.B., Lepow, M.L., McCracken, G.H., Peter, G., Phillips, C.F., Wright, H.T. and Plotkin, S.A., 1988, Haemophilus influenzae type b vaccines: lessons from the past, Pediatrics, 81:893.

Einhorn, M.S., Weinberg, G.A., Anderson, E.L., Granoff, P.D. and Granoff, D.M., 1986, Immunogenicity in infants of Haemophilus influenzae type b polysaccharide in a conjugate vaccine with Neisseria meningitidis outer-membrane protein, Lancet, ii:299.

Eskola, J., Peltola, H., Takala, A.H., Kayhty, H., Hakulinen, M., Karanko, V., Kela, E., Rekola, P., Ronnberg, P-R., Samuelson, J.S., Gordon, L.K. and Makela, P.H., 1987, Efficacy of Haemophilus influenzae type b polysaccharide-diphtheria toxoid conjugate vaccine in infancy, N.Engl.J.Med., 317:717.

Eskola, J., Kayhty, H., Peltola, H., Karanko, V., Makela, P.H., Samuelson, J. and Gordon, L.K., 1985, Antibody levels achieved in infants by course of Haemophilus influenzae type b polysaccharide/diphtheria toxoid conjugate vaccine, Lancet, i:1184.

Ferreccio, C., Levine, M.M., Rodriguez, H. and Contreras, R., 1989, Comparative efficacy of two, three or four doses of Ty21a live oral typhoid vaccine in enteric-coated capsules. A field trial in an endemic area, J.Infect.Dis., 159:766.

Finne, J., Leinonen, M. and Makela, P.H., 1983, Antigenic similarities between brain components and bacteria causing meningitidis. Implications for vaccine development and pathogenesis, Lancet, ii:355.

Fothergill, L.D. and Wright, J., 1933, Influenzal meningitis: relation of age incidence to the bactericidal power of blood against the causal organisms, J.Immunol., 24:273.

Gold, R.C., Lepow, M.L., Goldschneider, I., Draper, T.F. and Gotschlich, E.C., 1975, Clinical evaluation of group A and group C meningococcal polysaccharide vaccines in infants, J.Clin.Invest., 56:1536.

Gordon, L.K., 1984, Characterization of a hapten-carrier conjugate vaccine: H. influenzae-diphtheria conjugate vaccine, in Chanock, R.M. and Lerner, R.A., eds., Modern Approaches to Vaccines, Cold Spring Harbor Laboratory, Cold Spring Harbor, New York.

Gotschlich, E.C., Goldschneider, I. and Artenstein, M.S., 1969a, Human immunity to the meningococcus. V. The effect of immunization with the meningococcal group C polysaccharide on the carrier state, J.Exp.Med., 129:1385.

Gotschlich, E.C., Liu, T.Y. and Artenstein, M.S., 1969b, Human immunity to the meningococcus. III. Preparation and immunochemical properties of the group A, group B and group C meningococcal polysaccharide, J.Exp.Med., 129:1349.

Granoff, D.M. and Basden, M., 1980, Hemophilus influenzae infections in Fresno County, California: a prospective study of the effects of age, race and contact with a case on incidence of disease, J.Infect.Dis., 141:40.

Greenwood, B.M., Hassan-King, M., Whittle, H.C., 1978, Prevention of secondary cases of meningococcal disease in household contacts by vaccination, Br.Med.J., 1:1317.

Greenwood, B.M. and Wali, S.S., 1980, Control of meningococcal infection in
     the African meningitis belt by selective vaccination, Lancet, i:729.
Griffiss, J.M., Apicella, M.A., Greenwood, B. and Makela, P.H., 1987,
     Vaccines against encapsulated bacteria: a global agenda,
     R.Infect.Dis., 9:176.
Herva, E., Luotonen, J., Timonen, M., Sibakov, M., Karma, P. and Makela,
     P.H., 1980, The effect of polyvalent pneumococcal polysaccharide
     vaccine on nasopharyngeal and nasal carriage of streptococcus
     pneumoniae, Scand.J.Infect.Dis., 12:97.
Institute of Medicine, ed., 1986, New Vaccine Development, Establishing
     Priorities, Volume II, Diseases of importance in developing
     countries, Washington, D.C., National Academy Press.
Jann, K. and Jann, B., 1985, Cell surface components and virulence:
     Escherichia coli O and K antigens in relation to virulence and patho-
     genicity, in: The Virulence of Escherichia coli, Soc. Gen.Microbiol.
Jennings, H. and Lugowski, C., 1981, Immunochemistry of groups A, B and C
     meningococcal polysaccharide tetanus toxoid conjugates, J.Immunol.,
     127:104.
Karma, P., Pukander, J. and Sipila, M., 1985, Prevention of otitis media in
     children by pneumococcal vaccination, Am.J.Otolaryngol., 6:173.
Kayhty, H., Karanko, V., Peltola, H., Sarna, H. and Makela, P.H., 1980,
     Serum antibodies to capsular polysaccharide vaccine of group A
     Neisseria meningitidis followed for three years in infants and
     children, J.Infect.Dis., 142:861.
Kayhty, H., Peltola, H., Karanko, V. and Makela, P.H., 1983, The protective
     level of serum antibodies to the capsular polysaccharide of
     Haemophilus influenzae type b, J.Infect.Dis., 147:1100.
Kayhty, H., Karanko, V., Peltola, H and Makela, P.H., 1984, Serum antibodies
     after vaccination with Haemophilus influenzae type b capsular poly-
     saccharide and responses to reimmunization: no evidence of immuno-
     logic tolerance or memory, Pediatrics, 74:857.
Kayhty, H., Eskola, J., Peltola, H., Stout, M.G., Samuelson, J.S. and
     Gordon, L.K., 1987, Immunogenicity in infants of Haemophilus
     influenzae type b capsular polysaccharide mixed with DTP or conjug-
     ated to diphtheria toxoid, J.Infect.Dis., 155:100.
Kayhty, H., Makela, O., Eskola, J., Saarinen, L. and Seppala, I., 1988,
     Isotype distribution and bacteridal activity of antibodies after
     immunization with Haemophilus influenzae type b vaccines at 18-24
     months of age, J.Infect.Dis., 158:973.,
Kayhty, H., Peltola, H., Eskola, J., Ronnberg, P-R., Kela, E., Karanko, V.
     and Makela, P.H., 1989, Immunogenicity of Haemophilus influenzae
     oligosaccharide-protein and polysaccharide-protein conjugate vaccin-
     ation of children at 4, 6 and 14 months of age, Pediatrics, 84:995.
King, S.D., Ramlal, A., Wynter, H., Moodie, K., Castle, D., Kuo, J.S.C.,
     Barnes, L. and Williams, C.L., 1981, Safety and immunogenicity of a
     new Haemophilus influenzae type b vaccine in infants under one year
     of age, Lancet, ii:705.
Klein, J.O., 1981, The epidemiology of pneumococcal disease in infants and
     children, Rev.Infect.Dis., 3:246.
Koskela, M., 1986, Antibody response of young children to parenteral vaccin-
     ation with pneumococcal capsular polysaccharides: a comparison
     between antibody levels in serum and middle ear effusion,
     Pediatr.Infect.Dis., 5:431.
Lapeyssonnie, L., 1963, La meningite cerebro-spinale en Afrique,
     Bull.Wld.Hlth.Org., 28, Suppl.
Leinonen, M., Frasch, C.E., 1982, Class-specific antibody response to group
     B Neisseria meningitidis capsular polysaccharide: Use of polylysine
     precoating in an enzyme-linked immunosorbent assay, Infect.Immn.,
     38:1203.
Leinonen, M., Sakkinen, A., Kalliokoski, R., Luotonen, J., Timonen, M.,
     Makela, P.H., 1986, Antibody response to 14-valent pneumococcal

capsular polysaccharide vaccine in pre-school age children,
Pediatr.Infect.Dis., 5:39.

Levine, M.M., Ferreccio, C., Black, R.E. and Germanier, R., 1987a, Large-
scale field trial of Ty21a live oral typhoid vaccine in enteric-
coated capsule formulation, Lancet, i:1049.

Levine, M.M., Herrington, D., Murphy, J.R., Morris, J.G., Losonsky, G.,
Tall, B., Lindberg, A.A., Svenson, S., Bagar, S., Edwards, M.F. and
Stocker, B., 1987b, Safety, infectivity, immunogenicity and in vivo
stability of two attenuated auxotrophic mutant strains of Salmonella
typhi, 541Ty and 543Ty, as live oral vaccines in man, J.Clin.Invest.,
79:888.

Lindberg, A.A., Karnell, A., Pap, T., Sweiha, H., Hultenby, K. and Stocker,
B., 1990, Construction of an auxotrophic Shigella flexneri strain
for use as a live vaccine, M.Pathogen., 8, in press.

Losonsky, G.A., Santosham, M., Sehgal, V.M., Zwahlen, A. and Moxon, R.,
1984, Haemophilus influenzae disease in the White Mountain Apaches:
Molecular epidemiology of a high risk population,
Pediatr.Infect.Dis., 3:539.

Makela, O., Mattila, P., Rautonen, N., Seppala, I., Eskola, J. and Kayhty,
H., 1987, Isotype concentrations of human antibodies to Haemophilus
influenzae type b polysaccharide (Hib) in young adults immunized
with the polysaccharide as such or conjugated to a protein
(diphtheria toxoid), J.Immunol., 139:1999.

Michaels, R.H., Poziviak, C.S., Stonebraker, F.E., Norden, C.W., 1976,
Factors affecting pharyngeal Haemophilus influenzae type b colonization rates in children, J.Clin.Microbiol., 4:413.

Paton, J.C., Lock, R.A. and Hansman, D.J., 1983, Effect of immunization with
pneumolysin on survival time of mice challenged with Streptococcus
pneumoniae, Infect.Immun., 40:548.

Peltola, H., 1987, Meningococcal disease: an old enemy in Scandinavia, in:
Evolution of Meningococcal Disease, vol. I, Vedros, N.A., ed.,
University of California, Berkeley.

Peltola, H., Kayhty, H., Sivonen, A. and Makela, P.H., 1977a, Haemophilus
influenzae type b capsular polysaccharide vaccine in children: a
double blind study of 100,000 vaccinees three months to five years
of age in Finland, Pediatr., 60:730.

Peltola, H., Makela, P.H., Kayhty, H., Jousimies, H., Herva, E., Hallstrom,
K., Sivonen, A., Renkonen, O-V., Pettay, O., Karanko, V., Ahvonen,
P. and Sarna, S., 1977b, Clinical efficacy of meningococcus group A
capsular polysaccharide vaccine in children three months to five
years of age, N.Engl.J.Med., 297:686.

Peltola, H., Kayhty, H., Virtanen, M. and Makela, P.H., 1984, Prevention of
Haemophilus influenzae type b bacteremic infections with the capsular
polysaccharide vaccine, N.Engl.J.Med., 310:1561.

Pichichero, M.E. and Insel, R.A., 1983, Mucosal antibody response to parenteral vaccination with Haemophilus influenzae type b capsule,
J.Allergy.Clin.Immunol., 72:481.

Reingold, A.L., Broome, C.V., Hightower, A.W., Ajello, G.W., Bolan, G.A.,
Adamsbaum, C., Jones, E.E., Phillips, C., Tiendrebeogo, H. and
Yada, A., 1985, Age specific differences in duration of clinical
protection after vaccination with meningococcal polysaccharide A
vaccine, Lancet, ii:114.

Riley, I.D., Everingham, F.A., Smith, D.E. and Douglas, R.M., 1981,
Immunization with a polyvalent pneumococcal vaccine. Effect of
respiratory mortality in children living in the New Guinea highlands,
Arch.Dis.Child., 56:354.

Robbins, A. and Freeman, P., 1988, Obstacles to developing vaccines for the
Third World, Sci.Am., 259:90.

Robbins, J.B., Parke, J.C., Schneerson, R. and Whisnant, J.K., 1973,
Quantitative measurement of "natural" and immunization-induced
Haemophilus influenzae type b capsular polysaccharide antibodies,
Pediatr., 7:103.

Robbins, J.B., Austrian, R., Lee, C-J., Rastogi, S.C., Schiffman, G., Henrichsen, J., Makela, P.H., Broome, C.V., Facklam, R.R., Tiesjema, R.H. and Parke, J.C., Jr., 1983, Considerations for formulating the second-generation pneumococcal capsular polysaccharide vaccine with emphasis on the cross-reactive types within groups, J.Infect.Dis., 148:1136.

Rosen, C., Christensen, P., Hovelius, B. and Prellner, K., 1984, A longitudinal study of the nasopharyngeal carriage of pneumococci as related to pneumococcal vaccination in children attending day care centres, Acta Otolaryngol. (Stockh.), 98:524.

Sarnesto, A., Ranta, S., Vaananen, P. and Makela, O., 1985, Proportions of Ig classes and subclasses in rubella antibodies, Scand.J.Immunol., 21:275.

Saukkonen, K., Haltia, M., Frosch, M., Bitter-Suerman, D. and Leinonen, M., 1986, Antibodies to the capsular polysaccharide of Neisseria meningitidis group B or E. coli K1 bind to the brains of infant rats in vitro but not in vivo, Microb.Path., 1:101.

Schneerson, R., Barrera, O., Sutton, A. and Robbins, J.B., 1980, Preparation, characterization and immunogenicity of Haemophilus influenzae type b polysaccharide-protein conjugates, J.Exper.Med., 152:361.

Seppala, I.J.T., Rautonen, N., Sarnesto, A., Mattila, P.A. and Makela, O., 1984, The percentages of six immunoglobulin isotypes in human antibodies to tetanus toxoid: standardization of isotype-specific second antibodies in solid-phase assay, Eur.J.Immunol., 14:868.

Seppala, I., Sarva, H., Makela, O., Mattila, P., Eskola, J. and Kayhty, H., 1988, Human antibody responses to two conjugate vaccines of Haemophilus influenzae type b saccharides and diphtheria toxin, Scand.J.Immunol., 28:471.

Shann, F., 1986, Etiology of severe pneumonia in children in developing countries, Pediatr.Infect.Dis., 5:247.

Shann, F., Germer, S., Hazlett, D., Gratter, M., Linneman, V. and Payne, R., 1984, Aetiology of pneumonia in children in Goroka Hospital, Papua, New Guinea, Lancet, ii:537.

Szu, S.C., Stone, A.L., Robbins, J.D., Schneerson, R. and Robbins, J.B., 1987, Preparation and characterization of conjugates of the Vi capsular polysaccharide and carrier proteins, J.exp.Med., 166:1510.

Szu, S.C., Li, X., Schneerson, R., Vickers, J.H., Bryla, D. and Robbins, J.B., 1989, Comparative immunogenicities of Vi polysaccharide-protein conjugates composed of cholera toxin or tis b subunit as a carrier bound to high- or lower-molecular-weight Vi, Infect.Immun., 57:3823.

Takala, A.K., 1989, Epidemiologic characteristics and risk factors for invasive Haemophilus influenzae type b disease in a population with high vaccine efficacy, Pediatr.Infect.Dis., 8:343.

Takala, A.K., Eskola, J., Peltola, H. and Makela, P.H., 1989, Epidemiology of invasive Haemophilus influenzae type b disease among children in Finland before vaccination with Haemophilus influenzae type b conjugate vaccine, Pediatr.Infect.Dis.J., 8:297.

Takala, A.K., Eskola, J., Nissinen, A., Leinonen, M., Pekkanen, E. and Makela, P.H., 1990, Effect of vaccination with Haemophilus influenzae type b (Hib) conjugate vaccine on the oropharyngeal carriage of Hib, 30th ICAAC, Atlanta, Ga.

Todd, J.K. and Bruhn, F.W., 1975, Severe H. influenzae infections: spectrum of disease, Am.J.Dis.Child., 129:607.

Wahdan, M.H., Serie, C., Cerisier, Y., Sallam, S. and Germanier, R., 1982, A controlled field trial of live Salmonella typhi strain Ty21a oral vaccine against typhoid: Three years' results, J.Infect.Dis., 145:292.

Ward, H. and Cochi, S.L., 1988, Haemophilus influenzae vaccines, in: Vaccines, Plotkin, S.A., Mortimer, E.A., eds., W.B. Saunders Co., Philadelphia.

Ward, H., Margolis, H.S., Lum, M.K.W., Fraser, D.W., Bender, T.R. and
    Anderson, P., 1981, Haemophilus influenzae disease in Alaskan
    eskimos: Characteristics of a population with an unusual incidence
    of invasive disease, Lancet, i:1281.
Ward, J.I., Broome, C.V., Harrison, L.H., Shinefield, H. and Black, S.,
    1988, Haemophilus influenzae type b vaccines: lessons for the future,
    Pediatrics, 81:887.
Ward, J., Brenneman, G., Letson, W. and Heyward, W.L., 1990, Alaska H.
    influenzae vaccine study group. Limited efficacy of an Haemophilus
    influenzae type b conjugate vaccine (PRP-D) in Alaskan native infants
    immunized at 2, 4 and 6 months of age, N.Engl.J.Med., in press.

CURRENT PROGRESS AND FUTURE TRENDS IN BIRTH CONTROL VACCINES

Norbert Dreifurst and Avrion Mitchison

Deutsches Rheuma Forschungszentrum Berlin
Am Kleinen Wannsee 5
D-1000 Berlin-5, Germany

INTRODUCTION

This paper deals with birth control vaccines and within that area it concentrates on an important issue: the future of gonadotrophin vaccines. This choice has been made on two grounds. Although many different types of birth control vaccine are in principle possible, gonadotrophin-based ones are at the present time far the most advanced. Indeed two such vaccines have already completed their first clinical trials, and having been judged safe are now ready to be tested for efficacy. Our second reason for making this choice is personal experience. One of us currently serves as Chairman of the Vaccines Task Force of the World Health Organization's Human Reproduction Programme; in addition he can claim some familiarity with the gonadotrophin vaccine being developed at the National Institute of Immunology at New Delhi, largely through old friendship with Professor G.P. Talwar, its Director.

The concept of a birth control vaccine is simple, appealing, and at the same time somewhat paradoxical. The idea that one could take a component of the reproductive system and use it as a vaccine so as to raise a contraceptive level of immunity is simple enough and appeals to policy makers on the grounds that vaccines tend to be cheap, safe, acceptable and effective. This audience does not need to be reminded that vaccinology is still basking in the glory of the eradication of smallpox and will be familiar with the widely held view that the move from first to second generation vaccines holds great promise. The paradox is that such vaccines are based on normal products of the body: how then can the immune system ever regard them as foreign, and if it does succeed in doing so, how can the threat of auto-immune disease ever be alleviated?

The trick is to identify those components of the reproductive system that can safely be made the target of immunological attack. Such components will be more or less unique to this system and not present elsewhere in the body. The qualifier is included here because it has been surprisingly hard to find components which are absolutely unique: again and again what looks like the perfect antigen has been identified, but on further scrutiny turns out to be expressed in non-reproductive tissue as well. The effect of this has been to make investigators reluctant to exclude candidate vaccines simply because they are expressed elsewhere weakly, i.e. in very small quantities or through an immunological cross-reaction. The second important

Vaccines, Edited by G. Gregoriadis *et al.*
Plenum Press, New York, 1991

qualification for use in a vaccine is that the component should be dispensable, so that its elimination by immunological attack does not prejudice the survival of the reproductive system as a whole. A third qualification is that the component should be some well defined protein or peptide, susceptible to cheap manufacture by bio-engineering or by chemical synthesis. The possibility of engineering a vaccine-gene into a live virus vector can be considered for this purpose, although this raises obvious problems of safety. For the time being this criterion of ease of manufacture probably excludes the use of carbohydrates, even though some of the mucuses and jellies that abound in the reproductive system look tempting from the vaccine point of view.

There is a certain immunological logic in picking an antigen that is unique to the reproductive system, as this makes it less likely to have induced immunological tolerance of self and thereby excluded itself from immunological attack. Even so it is surprising for most immunologists to find proteins expressed on the trophoblast surface being taken seriously as vaccine candidates, but such is indeed the case. It all goes to show that an ounce of practice is worth a pound of theory and that nothing should be excluded on theoretical grounds alone.

So at the end of all this discussion, what serious candidates have emerged so far? Briefly, in rough order of priority, they can be listed as (1) chorionic gonadotrophin (2) spermatozoa surface glycoproteins (3) trophoblast surface glycoproteins, and (4) zona pelucida glycoproteins. This list could be somewhat extended by including candidates which are still poorly defined, such as certain proteins involved in transport into the embryo and those which most, but not all, vaccinologists deem too hazardous, such as pituitary releasing factors or testicular proteins. All of these candidates have attracted the interest of a fair number of investigators. Rather than review this work in detail, we refer to two recent publications each of which covers much of the literature (Mitchison, 1990a, b). After a brief comment on sperm vaccines, the remainder of this paper deals with the item placed at the head of this list, gonadotrophin-based vaccines.

SPERMATOZOA VACCINES

Spermatozoa-based vaccines make good sense for several reasons. In the female at least, sperm are located topologically outside the body, and so appropriate immunization with antigens from sperm could be expected to provoke a contraceptive response without otherwise harmful effects. Furthermore, circumstantial evidence indicates that immunity to sperm does occasionally cause sterility naturally and there is even some evidence that immunosuppression can alleviate this form of sterility, although a controlled trial has yet to be carried out. In contrast to most other possible targets of immunological attack, sperm must have come under only weak evolutionary pressure to reduce immunogenicity.

In spite of these attractions and a fair amount of research, progress with sperm-based vaccines has lagged. In part this is because there are too many, rather than too few, candidate vaccine molecules on the surface of sperm. Typically, investigators with a primary interest in sperm biology start raising monoclonal antibodies and in no time find themselves with a candidate in hand. There is little agreement about how large a pool is being fished, nor what constitutes an exclusion criterion able to eliminate unsuccessful candidates. For that reason the WHO Vaccines Task Force has tried to initiate an alternative gene-based approach. The idea is to screen a testis gene bank by expression, probably on the surface of COS cells. Screening could be by means of polyclonal antisera, or adhesion to, say, zona pelucida, or even, like Sutcliffe in Glasgow, by screening for all proteins with a membrane insert. The aim is to devise a method of rapidly identifying a large number of surface proteins, any of which could provisionally be regarded as a vaccine candidate. Only then need the arduous task of comparative evaluation be undertaken. But at least the evaluation would be carried out with all the entrants starting together: what point is there in running a race with just one horse at a time?

GONADOTROPHIN-BASED VACCINES

These vaccines have several advantages. The hormone is essential for the establishment of pregnancy, during which it is made by the trophoblast and acts on the corpus luteum to stimulate production of progesterone. It is made in substantial amounts only during pregnancy, so that neutralizing it should have no consequences other than to inhibit pregnancy. Early on it is made in quantities small enough to be neutralized easily by antibodies. These considerations have attracted the interest of several vaccine development groups and by now two broadly similar versions of this vaccine are well advanced. Both involve only the β chain of this two-chain hormone, as the α chain is also present in several other hormones that are essential for normal life.

One of these versions consists of the carboxy-terminal 37 amino acids of the β chain coupled as a peptide to diphtheria toxoid. This antigen, plus N-acetyl-nor muramyl-L alanyl-D isoglutamine as adjuvant, is administered as a water-in-oil emulsion in squalene plus mannide mono-oleate. This vaccine has been used successfully in a phase I clinical trial in Australia, where potentially contraceptive levels of antibody were maintained for as long as six months in the highest dose group (Stevens, 1981a, b; Jones et al, 1988).

In the other version the entire 145 amino acid β chain is conjugated to tetanus toxoid and administered on alum. An alternative formulation in which this chain is annealed to the α chain of ovine luteinizing hormone prior to conjugation, in order to preserve conformation, has also been tested. These vaccines have been used successfully in a phase I clinical trial in India, where potentially contraceptive levels of antibody were again maintained for about six months (Talwar and Raghupathy, 1989; Singh et al, 1989).

Comparing the two versions, the peptide vaccine has the merit that the peptide is available from chemical synthesis. On the other hand, the intact-chain vaccine is more immunogenic and therefore requires a less drastic adjuvant. What was originally regarded as a serious drawback, its cross reactivity with luteinizing hormone, now seems less serious in the light of the uninterrupted menstrual cycles observed in the subjects of the phase I trial, although the long term sequelae of this crossreaction remains a serious consideration. In fact both vaccines have one potential side effect in common, as a hormone is released in small amounts from the pituitary of non-pregnant women which has the same structure as chorionic gonadotrophin.

THE LESSON FOR OTHER VACCINES

This experience with gonadotrophin has useful lessons for the development of other vaccines, some scientific and others administrative. The scientific value of this experience stems from the fact that we are dealing with a self-protein and something which is therefore not intrinsically immunogenic. Even when the various manoeuvres described above have been performed, the antigens remain rather weak by the standards of most microbial vaccines and this has necessitated a great deal of study of carriers, adjuvants and delivery vehicles. This is particularly true of the peptide-based vaccine, where these studies have been both elaborate and expensive. As a result, an extensive knowledge-base on this topic has been acquired, which now is probably wider than that available for any other vaccine. That is all the more true because of the parallel development of the two forms of the vaccine, so that the scientific findings complement and reinforce one another.

It is therefore regrettable that one useful element in this knowledge-base seems not to be transferrable. That is the formulation of the present-ly preferred delivery system for the peptide vaccine, namely microcapsules devised and manufactured by the Stolle Corporation of Cincinatti. These provide an excellent form of controlled, long term delivery; the problem is that each new vaccine protein as it comes along seems to need reformulation, as the detailed make-up has to vary from one protein to another, or so it is claimed. All this is a little difficult to evaluate as the methods of en-capsulation employed by Stolle and other companies are shrouded in commer-cial secrecy.

The bottom line is that at the present moment the preferred adjuvant for the peptide vaccine is not very different from Freund's complete adjuv-ant, for long the gold standard in animal immunization but believed to be dangerous in man. The vaccine used in the first human trial, as mentioned above, was a water-in-oil emulsion made up in arlacel plus squalene, and incorporated a muramyl dipeptide derivative. The next trial is likely to use a similar emulsion, but this time with micro-encapsulated antigen within the aqueous phase.

A NOTE ON ADMINISTRATION

At first sight, the administrative structures behind the two vaccines look very different. On the one hand, the intact beta chain program is run by the National Institute of Immunology, New Delhi, the India Council for Medical Research and the Population Council in New York; on the other, the peptide program involves research in Ohio and Adelaide, combined with finan-cial support and advice from WHO. What the two programs have in common is that both have tiny but utterly committed leaderships: Pran Talwar and Rosemary Thau for the former program and Vernon Stevens, Warren Jones and David Griffin for the latter. Of course there are any number of advisory and assessment committees, but the similarity in achievement of the two programs suggests that what really matters is high-grade management, plus access to substantial funds. Small is beautiful.

The success of these programs contrasts sharply with the failure of investigator initiative, peer review structures to participate effectively in the middle and later stages of vaccine development. Where that sort of structure performs brilliantly is in the very early stages, when the ideas that can lead on to a managed program are first created and tested. In addition, that structure can function efficiently in designing and evalu-ating new technologies, such as the gene-based approach to sperm vaccines mentioned above. It is noticeable how easily investigators (and granting agencies outside WHO) get the two stages confused and claim to be "develop-ing a vaccine" when they have little idea of the scale of effort required and have no real intention of undertaking serious management tasks. In fact, the claim to be "developing a vaccine" has become something of a cliche in immunological circles.

This is where the hard thinking is needed. First, for how long will it make sense to go on running development of the two versions of this type of vaccine? Second, what further development will be needed before one or both vaccine can enter general use?

As regards the cost and benefit of running the two programs in parallel, my guess is that total expenditure must be running at a level of two to three million dollars per year and that that figure is likely to increase rather than decrease over the next five to ten years. That reflects the likely costs of safety testing, of moving from laboratory to manufacturing scale production and of large scale clinical trials. At the present moment, the costs of insurance in connection with future clinical trials loom large as a potential obstacle, at least so far as concerns the WHO program. Perhaps discontinuing one of the two programs would cut costs by half. Apart from cost, a good reason for making a decision sooner rather than later would be to minimize the exposure of women to unnecessary trial risks.

The benefit of continuing to run two programs, for the time being at least, can also be regarded in terms of minimizing risks. The worst risk of premature cancellation is that a sub-optimal vaccine might be chosen. Furthermore, it is entirely possible that the two vaccines may eventually find different niches in family planning: slightly different durations of effect, side effects, and costs are to be expected, for instance. Another obvious risk is that the chosen program might become sleepy without the competition. A less obvious risk is that a program based only in either a developing or a developed country will necessarily suffer certain constraints; by having both run, in parallel and with a high level of information exchange, these constraints can be minimized. They include, for instance, the different types of licensing and insurance requirements. This does not at all mean having lower safety standards in a developing country. To give a practical example, because of nutritional status and cultural background, trials in India usually cannot involve frequent collection of blood samples, although this may be mandatory in Germany.

In any case cancellation of either program in favour of the other would hardly be politically possible at the present time. Too many different agencies and nationalities are involved. Our view, as argued above, is that for the time being this is a very good thing, although in the long run the two vaccines wil surely need to be evaluated in competition with one another in the same trial. This is a task in which WHO should certainly be involved. In the meanwhile we can consider what further developments will be required in the two programs.

As regards the peptide vaccine, one immediate aim is to find a cleaner substitute for the ill-defined toxoid preparation that is at present used as the carrier. Thoughts turn naturally to a recombinant protein (i.e. one read off a cloned gene), but it is not really clear that this offers any significant advantage over a naturally occurring microbial protein. After all, the problems of purification will be much the same. The final choice will probably depend on a combination of second order factors, such as the proportion shunted into the culture medium, ease of bulk culture, yield, ease of purification, ease of coupling to the peptide, ease of incorporation into microcapsules, immunogenicity, and lack of side effects such as hypersensitivity. Even this list underestimates the problem, for studies with the alternative form of vaccine indicate that a carrier can be over- as well as under-immunogenic, probably because of intramolecular antigenic competition.

Most of the other pre-production problems have been solved. The program is well ahead with the chemistry of coupling, optimization of the peptide to carrier ratio, choice of adjuvant, and choice of delivery vehicle. But it must be remembered that these matters have been gone into so carefully precisely because this form of vaccine is so weakly immunogenic; were this not the case, such detailed optimization would not be needed. We still do not know whether even in its present form it will have an adequate contraceptive effect. Hence the need for an immediate phase two trial. In the meanwhile a search for other immunogenic peptides will continue, although obviously we do not want to re-invent the entire chain!

As regards the intact chain vaccine, some of the problems are similar to those just outlined, and others different. The problem of finding a carrier cleaner than the present toxoid is much the same, although the details of antigenic competition between carrier and peptide compared with those between carrier and intact chain are somewhat different. Here again, it is tempting to go for carrier and chain as a single engineered fusion protein. However, experience with the peptide conjugate suggests that about five chains per molecule of carrier would be optimal. That in turn indicates that chemical coupling may still be the best route to follow.

If one accepts this argument, then the next problem can be summarized thus: how would it be possible to obtain intact chain in quantities large enough and clean enough for a large scale manufacturing process? Pregnancy urine, whence supplies of the chain have so far been obtained, hardly seems feasible. So at this point we offer two highly dogmatic pieces of advice. Both of them are debatable even at present and could easily be falsified by some new turn of events. And you might well ask what business it is of ours anyway. Nevertheless we believe them to be true and important. The first is that at 145 amino acids, the beta chain is probably just small enough to be produced in soluble form by a prokaryote. Any larger, and problems are likely to be encountered of improper folding with consequent insolubility. One could reasonably expect production levels of the order of one gram protein per litre of culture.

Our second piece of advice is to go to one or more of the commercial companies with practical experience of that kind of production. We know full well that this is a matter where the costs and advantages need careful balancing. Our view is clear: so much has been learned by these companies that to try to go it alone at the present time would be extremely wasteful. As for the costs, consider the following thought. The first major call on the European Economic Community's science program in India was to spend around one milion Ecus on establishing a production facility for the new French polio vaccine. Surely a viable contraceptive vaccine offers an even more important goal: where better could the efforts of the Community be directed?

The question is often asked whether contraceptive vaccination is irreversible and therefore often unacceptable. The question of reversibility boils down to how long it would take for levels of antibody to fall below the effective level (in theory, pregnancy might be sustained by extrinsic progesterone even in the presence of high levels of antibody, but this possibility seems somewhat remote). At present the main problem confronting both of the gonadotrophin vaccine programs is not reversibility, but rather to sustain antibody levels for a sufficiently long period, over say two years. So far this has not been achieved and hence the interest in slow-release encapsulation systems. One can hope that in the future it may be possible to devise tailored release systems that would enable a woman to choose the period over which she would be protected against pregnancy. This kind of approach would also require some kind of dip-stick assay that could tell her whether her protection was still working.

THE MESSAGE FOR BIRTH CONTROL POLICIES

Birth control vaccines have reached a promising stage and every effort should be made to press on with their development and clinical testing. Not to do so might lose a valuable component in the response of science to the world's demographic crisis. At the same time it must be emphasised that not one of these vaccines has yet been proven effective in clinical trials. Accordingly, the promise that they hold should certainly not be taken as an excuse for reducing effort in improving and implementing other methods of birth control.

REFERENCES

Jones, W.R., Bradley, J., Judd, S.J., Denholm, E.H., Ing, R.M.Y., Muller, U.W., Powell, J., Griffin, T.D. and Stevens, V.C., 1988, Phase I clinical trials of the World Health Organization birth control vaccine, Lancet, i:1295.

Mitchison, N.A., 1990a, Gonadotrophin vaccines, J.Curr.Opin.in Immunol., in press.

Mitchison, N.A., 1990b, Status of immunocontraception and future needs, in: "Gamete Interation: Prospects for Immunocontrapception", N.J. Alexander, D. Griffin, J. Spieler and G. Waites, eds., Wylie Liss, N.Y., in press.

Singh, O.M., Rao, L.V., Gaur, A.M., Sharma, N.C., Ala, A. and Talwar, G.P., 1989, Antibody response and characteristics of antibodies in women immunized with three contraceptive vaccines inducing antibodies against human chorionic gonadotrophin, Fertil.Steril., 52:739.

Stevens, V.C., Cinader, B., Powell, J.E., Lee, A.C. and Kohn, S.W., 1981a, Preparation and formulation of a human chorionic gonadotrophin anti-fertility vaccine: selection of a peptide immunogen, Am.J.Reprod.Immunol., 1:307.

Stevens, V.C., Cinader, B., Powell, J.E., Lee, A.C. and Kohn, S.W., 1981b, Preparation and formulation of hCG anti-fertility vaccine: selection adjuvant and vehicle, Am.J.Reprod.Immunol., 1:315.

Talwar, G.P. and Raghupathy, R., 1989, Anti-fertility vaccines, Vaccine, 7:97.

CONTRIBUTORS

Allison, A.C., Syntex Research, Palo Alto, California, USA

Bartoloni, A., Sclavo Research Center, Via Fiorentina 1, 53100 Siena, Italy

Bhat, B., Wyeth-Ayerst Research, P.O. Box 8299, Philadelphia, PA 19101, USA

Bhat, R., Wyeth-Ayerst Research, P.O. Box 8299, Philadelphia, PA 19101, USA

Bomford, R., Wellcome Biotechnology, Langley Court, Beckenham, Kent BR3 3BS, UK

Bonde, G.M., Wyeth-Ayerst Research, P.O. Box 8299, Philadelphia, PA 19101, USA

Byars, N.E., Syntex Research, Palo Alto, California, USA

Chanda, P., Wyeth-Ayerst Research, P.O. Box 8299, Philadelphia, PA 19101, USA

Chengalvala, M., Wyeth-Ayerst Research, P.O. Box 8299, Philadelphia, PA 19101, USA

Covacci, A., Sclavo Research Center, Via Fiorentina 1, 53100 Siena, Italy

Davis, A., Wyeth-Ayerst Research, P.O. Box 8299, Philadelphia, PA 19101, USA

Dreifurst, N., Deutsches Rheuma Forschungszentrum Berlin, Am Kleinen Wannsee 5, D-1000 Berlin-5, Germany

Epstein, M.A., Nuffield Department of Clinical Medicine, University of Oxford, John Radcliffe Hospital, Oxford OX3 9DU, UK

Eskola, J., National Public Health Institute, SF 00300 Helsinki, Finland

Francis, M.J., Wellcome Research Laboratories, Langley Court, Beckenham, Kent BR3 3BS, UK

Goldberg, K.M., Wyeth-Ayerst Research, P.O. Box 8299, Philadelphia, PA 19101, USA

Harmsen, T., Eykman-Winkler Laboratory of Medical Microbiology, Utrecht University, Utrecht, The Netherlands

Heath, A.W., Paravax Inc., Mountain View, California, USA

Hite, M., Wyeth-Ayerst Research, P.O. Box 8299, Philadelphia, PA 19101, USA

Hjorth, R.N., Wyeth-Ayerst Research, P.O. Box 8299, Philadelphia, PA 19101, USA

Hsu, K-H., Wyeth-Ayerst Research, P.O. Box 8299, Philadelphia, PA 19101, USA

Hung, P.P., Wyeth-Ayerst Research, P.O. Box 8299, Philadelphia, PA 19101, USA

Kayhty, H., National Public Health Institute, SF 00300 Helsinki, Finland

Knight, S.C., Division of Immunological Medicine, MRC Clinical Research Centre, Watford Road, Harrow, Middx., HA1 3UJ, UK

Kostek, B., Wyeth-Ayerst Research, P.O. Box 8299, Philadelphia, PA 19101, USA

Kraaijeveld, C.A., Eykman-Winkler Laboratory of Medical Microbiology, Utrecht University, Utrecht, The Netherlands

Levner, M.H., Wyeth-Ayerst Research, P.O. Box 8299, Philadelphia, PA 19101, USA

Lubeck, M., Wyeth-Ayerst Research, P.O. Box 8299, Philadelphia, PA 19101, USA

Makela, P.H., National Public Health Institute, SF 00300 Helsinki, Finland

Mason, B., Wyeth-Ayerst Research, P.O. Box 8299, Philadelphia, PA 19101, USA

Mitchison, A., Deutsches Rheuma Forschungszentrum Berlin, Am Kleinen Wannsee 5, D-1000 Berlin-5, Germany

Mizutani, S., Wyeth-Ayerst Research, P.O. Box 8299, Philadelphia, PA 19101, USA

Nakano, G.M., Syntex Research, Palo Alto, California, USA

Natuk, R., Wyeth-Ayerst Research, P.O. Box 8299, Philadelphia, PA 19101, USA

Nencioni, L., Sclavo Research Center, Via Fiorentina 1, 53100 Siena, Italy

Nucci, D., Sclavo Research Center, Via Fiorentina 1, 53100 Siena, Italy

Oosterlaken, T.A.M., Eykman-Winkler Laboratory of Medical Microbiology, Utrecht University, Utrecht, The Netherlands

Peltola, H., National Public Health Institute, SF 00300 Helsinki, Finland

Piner, E.D., Wyeth-Ayerst Research, P.O. Box 8299, Philadelphia, PA 19101, USA

Pizza, M.G., Sclavo Research Center, Via Fiorentina 1, 53100 Siena, Italy

Playfair, J.H.L., Department of Immunology, UCMSM, London, UK

Rappuoli, R., Sclavo Research Center, Via Fiorentina 1, 53100 Siena, Italy

Selling, B., Wyeth-Ayerst Research, P.O. Box 8299, Philadelphia, PA 19101, USA

Snippe, H., Eykman-Winkler Laboratory of Medical Microbiology, Utrecht University, Utrecht, The Netherlands

Stapleton, M., Wellcome Biotechnology, Langley Court, Beckenham, Kent
BR3 3BS, UK

Stewart-Tull, D.E.S., Department of Microbiology, University of Glasgow,
Glasgow G12 8QQ, UK

Takala, A.K., National Public Health Institute, SF 00300, Helsinki, Finland

Teller, D.M., Wyeth-Ayerst Research, P.O. Box 8299, Philadelphia, PA 19101,
USA

Verheul, A.F.M., Eykman-Winkler Laboratory of Medical Microbiology, Utrecht
University, Utrecht, The Netherlands

Vernon, S.K., Wyeth-Ayerst Research, P.O. Box 8299, Philadelphia, PA 19101,
USA

Virelizier, J-L., Unite d'Immunologie Virale, Institut Pasteur 75724, Paris,
Cedex 15, France

Welch, M., Syntex Research, Palo Alto, California, USA

Winsor, S., Wellcome Biotechnology, Langley Court, Beckenham, Kent BR3 3BS,
UK

Zigterman, G.J.W.J., Department of Bacteriology Research, Intervet
International, Boxmeer, The Netherlands

Participants of the NATO Advanced Studies Institute "Vaccines: Recent Trends and Progress" held at Cape Sounion, Greece during 24 June–5 July, 1990. The organizing committee included Anthony C. Allison (ASI Co-director), K. Dalsgaard, Gregory Gregoriadis (ASI Director and Chairman), George Poste (ASI Co-director) and Harm Snippe.

INDEX

153